Move Off
The Grid

Move Off The Grid

K. K. Yadhunath

PARTRIDGE
A Penguin Company

Copyright © 2013 by K.K.Yadhunath.

ISBN: Softcover   978-1-4828-0087-6
          Ebook   978-1-4828-0086-9

All rights reserved. No part of this book may be used or reproduced by any means, graphic, electronic, or mechanical, including photocopying, recording, taping or by any information storage retrieval system without the written permission of the publisher except in the case of brief quotations embodied in critical articles and reviews.

Because of the dynamic nature of the Internet, any web addresses or links contained in this book may have changed since publication and may no longer be valid. The views expressed in this work are solely those of the author and do not necessarily reflect the views of the publisher, and the publisher hereby disclaims any responsibility for them.

This preclude the information directly adopted as fair use from various organization's websites
Photo Credit:
Front cover—Gray Watson—Boston
Back Cover—Kathy Crane—NOAA Artic Research program

Edition: EBook2013-A

Information contained in this work has been obtained by the Publisher / Author, from sources believed to be reliable. However neither the publisher nor the author guarantees the accuracy or completeness of any information published herein and neither Publisher nor the author shall be responsible for any errors, omission, or damages arising out of use of this information. This work is published with the understanding that Publisher and the author are supplying information but not rendering design, engineering or other professional services. If such services are required, the assistance of an appropriate professional should be sought.

K.K.Yadhunath
yadhupost@gmail.com

Printed in India.

Partridge books may be ordered through booksellers or by contacting:

Partridge India
Penguin Books India Pvt.Ltd
11, Community Centre, Panchsheel Park, New Delhi 110017
India
www.partridgepublishing.com
Phone: 000.800.10062.62

"Dedicated to the nation's journey towards independent power"

"O Mother Earth! You are the world for us and we are your children; let us speak in one accord; let us come together so that we live in peace and harmony!"

Atharva Veda

CONTENTS

Sl No	Chapter	
1	History & Evolution of Solar	1
2	Solar Energy Harnessing—How it works	13
3	Solar Technologies Explained	24
4	Solar Energy Products	55
5	Climate Change	84
6	Solar Energy—Global Scenario	108
7	India Electric Power scenario	137
8	Solar Power In India	157
9	National Solar Mission	187
10	Move Off The Grid	201
11	How to size a SHS & SWH	216
12	UAE Green City	229
13	Energy Efficiency	238
14	Other Renewable Technologies	258
15	Green Buildings	292
16	Electric Vehicles	302

Add on Info

1	A list of 100 Solar Products	314
2	Solar power plants around the world:	316
3	A list of Solar companies in India	319
4	A List of Banks financing for Solar	321
5	Domestic appliances: Power	322
6	Units and Conversion Factors:	323
7	Calorific Value	325

PREFACE

This simple book is about the use of solar energy at our home. The chapters talk about the renewable energy and its importance for a sustainable living. The amazing journey of solar power from being one providing only daylight to be an alternate source for energy to mankind and its life saving impact is scripted here through different chapters. Our scientists have tirelessly worked to bring solar to its current form of ready to use packaged power. It is a naked truth today that the journey of harnessing solar energy has made substantial achievements. But as it always has been, much more of sun and the use of its energy remained unexplored of what science have achieved so far.

Solar energy is now the cleanest form of energy. Even though it remained the most difficult to harness among renewables, solar energy has now come off age and poised to compete directly with grid electricity in this decade. Worldwide solar energy is seeing exponential growth with support from the Investors, Industry, Governments, Environmentalists and Customers among others. There are many established and evolving technologies to harness solar power.

Solar power installations are currently the fastest growing among all power generation installations. With a stupendous growth worldwide, Solar in India is aligning itself for a revolutionary growth in installation and usage. Since the seventies the Government of India has supported and encouraged utilization of solar energy. Now we have the ambitious programme of JNSM with a mission target of 20GW by 2022.

A few chapters in this book may contain an excess of numbers and graphical representations. Some other chapters provide technical details. This information is here to emphasize the fact that—Solar will work and is working around the world. I have tried to make these as simple as possible for all discerning readers to easily understand.

The introductory chapter of the book looks into the history of solar energy. The next couple of chapters get more specific on solar energy and its usage. The amazing range of products that work on solar which gives us an alternate to grid power is briefed here next.

The next session of the book talks about climate change, its impact on our ecology and an exhaustive list of organizations working to combat climate change. This is followed by the solar energy utilisation around the world.

India's power requirements is then followed by the status of solar power in India, The aggressive action plan of the government of India, a serious look on why to minimize the use of grid power and then the tips for selecting and installing the right solar system for your use.

The wonderful concept of world's first carbon neutral city comes next followed by a chapter on the importance of energy efficiency. The world of renewables has an astonishing range of power generation techniques and this is detailed in the next chapter. This is followed by the chapter on Green buildings. The main book comes to a landing with a chapter on one of the most promising concept—Electric Vehicles.

The centurial range of solar products, a list of solar power plants around the world, the current stake holders of solar in India, the banks supporting solar initiative through various credit schemes are all listed. A few details on the power of household equipments, useful units of power culminate in completing this book on contemporary science and its imminent use.

I had thought for sometime now that the subject of solar energy use and sustainable living needs more explanation which will help in popularizing solar energy utilization to the common man. This was reinforced while I set out to sell the solar products for retail homes, more so when I was encountered with a variety of questions. This book is an attempt in this regard.

The preface is kept very brief for you to read the book fast, enjoy and understand about sustainable living by having your own power generation at home.

Happy reading!

K.K.Yadhunath

ACKNOWLEDGEMENTS

I thank all the people who provided valuable inputs during the preparation of this book. This book contains specifics related to the usage, current development & growth of renewable energy, solar in particular. Primarily making a serious effort to convey that solar energy harnessing is a reality now and its adoption will usher in an era of sustainable living.

I acknowledge with thanks the reference of publications in the public domain from World Bank, United Nations, IPCC, NREL, IEA, IRENA, MNRE, Wikipedia etc where I could research different facets of energy and its usage among others. Further wish to add that this work is in no way related to the above or any other organisations nor they endorse this book.

I thank my colleagues, staff and senior management of Orb energy and Shell Solar Lanka where I am working/ worked which helped me in having a first hand information about the dissemination of solar energy.

I thank my family for their patience and support during the preparation the book.

Green energy activists who have been promoting the use of renewables have always been a source of inspiration for me which is duly acknowledged here.

K.K.Yadhunath

CHAPTER 1

HISTORY & EVOLUTION OF SOLAR

> *God forbid that India should ever take to industrialism after the manner of the west . . . keeping the world in chains. If [our nation] took to similar economic exploitation, it would strip the world bare like locusts.*
>
> —*Mahatma Gandhi*

For billions of years, the power of the Sun has been harnessed by plants through photosynthesis to produce glucose and cellulose. A by product of this process, is oxygen, a substance required by all living beings on Earth.

The Vedic scriptures of the Hindu religion refer to the sun as the store house of inexhaustible power and radiance. The sun god is referred to as Surya or Aditya. The Vedas are full of hymns describing the celestial body as the source and sustainer of all life on earth. The worship of the Sun in India is several centuries old.

The ancient Egyptians used solar heat to bake mixtures of mud and straw into hard bricks for construction. The Greeks and Romans used clever architecture to make use of the Sun's ability to provide light and heat in a building.

In 200BC the Greeks used bronze shields to focus sun rays on Roman ships, setting them on fire. In 20 B.C Chinese used sun rays reflected from shields to light torches for religious services.

During ancient times the Greeks, the Chinese and the Native Americans were using the sun to warm their homes and keep them disease free.

Photovoltaic Solar energy:

The invention of Photovoltaic solar energy happened in the early period of 19th century and experiments to bring it to a usable form continued at different levels thereon.

Period: 1830-1900

The most important incident in solar energy history was the discovery of photovoltaic effect in 1839 AD. It was done by the famous French physicist Edmund Becquerel. He found that some materials produce electric current when they are exposed to sun light.

- 1873—Willoughby Smith discovered the photoconductivity of selenium.
- 1876—William Grylls Adams and Richard Evans Day discovered that Selenium produces electricity when exposed to light. Although selenium solar cells failed to convert enough sunlight to power electrical equipment, they proved that a solid material could change light into electricity without heat or moving parts
- 1883—Charles Fritts, an American inventor, described the first solar cells made from selenium wafers.
- 1887—Heinrich Hertz discovered that ultraviolet light altered the lowest voltage capable of causing a spark to jump between two metal electrodes.

Period: 1900-1945

- 1905—Albert Einstein published his paper on the photoelectric effect (along with a paper on his theory of relativity).
- 1916—Robert Millikan provided experimental proof of the photoelectric effect.
- 1918—Polish scientist Jan Czochralski developed a way to grow single—Crystal Silicon
- 1921—Albert Einstein wins the Nobel Prize for his theories (1904 research and technical paper) explaining the photoelectric effect.

Period: 1945-1970

1954—Photovoltaic technology is born in the United States when Daryl Chapin, Calvin Fuller, and Gerald Pearson developed the silicon photovoltaic (PV) cell at Bell Labs—the first solar cell capable of converting enough of the sun's energy into power to run everyday electrical equipment. Bell Telephone Laboratories produced silicon solar cell with 4% efficiency and later achieved 11% efficiency.

1955—Western Electric began to sell commercial licenses for silicon photovoltaic (PV) technologies. Early successful products included PV-powered dollar bill changers and devices that decoded computer punch cards and tape.

1958—The solar cells were applied to satellite systems. The solar cells provided electrical power to a satellite Vanguard 1 for 6 years after which the satellite was decommissioned. This marked the beginning of a realization that solar energy could be looked at as a renewable source.

—In Japan, the country's first PV system of capacity 70w was installed for a radio station on Mount Shinobuyama

1959—Hoffman Electronics achieves 10% efficient with commercially available photovoltaic cells. Hoffman also learns to use a grid contact, reducing the series resistance significantly.

On August 7, the Explorer VI satellite was launched with a photovoltaic array to recharge the batteries while in orbit. On October 13, the Explorer VII satellite was launched powered by 3000 solar cells.

—Sharp Japan, commence development of solar cells

1960—Hoffman Electronics achieves 14% efficient photovoltaic cells.

1963—Japan installs a 242-watt, photovoltaic array on a lighthouse, the world's largest array at that time.

1966—NASA launches the first Orbiting Astronomical Observatory, powered by a 1-kilowatt photovoltaic array, to provide astronomical data in the ultraviolet and X-ray wavelengths filtered out by the earth's atmosphere.

Period: 1970-2000

1970—The technology existed but was too expensive to be applied for civilian setups. Dr. Elliot Berman designed the solar cell that had an easier and cheaper construction cost as compared to the super expensive solar power cells used in satellites. Dr. Elliot Berman, with help from Exxon Corporation, designs a significantly less costly solar cell, bringing price down from $100 a watt to $20 a watt. Solar cells began to power navigation warning lights and horns on many offshore gas and oil rigs, lighthouses, railroad crossings etc. During this decade domestic solar applications began to be viewed as sensible applications in remote locations where grid connected utilities could not exist affordably. At this point, the complete solar energy history took a new course towards higher, thinner and cheaper solar cell technology.

1972—The French installed a cadmium sulfide (CdS) photovoltaic system to operate an educational television at a village school in Niger.

—The Institute of Energy Conversion was established at the University of Delaware to perform research and development on thin-film photovoltaic (PV) and solar thermal systems, becoming the world's first laboratory dedicated to PV research and development.

1973—The University of Delaware builds "Solar One," one of the world's first photovoltaic (PV) powered residences. The system is a PV/thermal hybrid. The roof-integrated arrays fed surplus power through a special meter to the utility during the day and purchased power from the utility at night.

1975—Kyocera establish the Japan Solar Energy Corp (JSEC) to engage in R&D of solar cells

1976—Dr. David Carlson and Dr. Christopher Wronski co-invented the hydrogenated amorphous silicon solar cell.

—The NASA Lewis Research Center starts installing 83 photovoltaic power systems on every continent except Australia. These systems provide such diverse applications as vaccine refrigeration, room lighting, medical clinic lighting, telecommunications, water pumping, grain milling, and classroom television. The Center completed the project in

1995, working on it from 1976-1985 and then again from 1992-1995.

—The U.S. Department of Energy launches the Solar Energy Research Institute (SERI) now NREL, http://www.nrel.gov/ "National Renewable Energy Laboratory", a federal facility dedicated to harnessing power from the sun.

1977—Total photovoltaic manufacturing production exceeds 500 kilowatts.

1978—USA with its National Energy Act implemented the first form of feed in tariff

1980—ARCO Solar becomes the first company to produce more than 1 megawatt of photovoltaic modules in one year.

—At the University of Delaware, the first thin-film solar cell exceeds 10% efficiency using copper sulfide/cadmium sulfide.

1985—The University of South Wales breaks the 20% efficiency barrier for Silicon solar cells under 1-sun conditions.

1986—ARCO Solar releases the G-4000—the world's first commercial thin—film power module.

1992—Thin film cadmium Telluride module with over 15% efficiency developed

1993—Pacific Gas & Electric completes installation of the first grid—supported photovoltaic system in Kerman, California. The 500—kilowatt system was the first "distributed power" effort.

1996—Solar powered airplane—Icare—one of the most advanced made its flight in Germany

1998—Subhendu Guha, a noted scientist for his pioneering work in amorphous silicon, led the invention of flexible solar shingles, a roofing material and state-of-the-art technology for converting sunlight to electricity.

1999—Construction was completed on 4 Times Square, the tallest skyscraper built in the 1990s in New York City. It incorporated more energy-efficient building techniques than any other commercial skyscraper and also included building-integrated photovoltaic (BIPV) panels on the 37th through 43rd floors on the southand west-facing facades that produce a portion of the buildings power.

1999—Cumulative worldwide installed photovoltaic capacity reaches 1000 megawatts

Period: 2000-2005

2000—First Solar started production in Perrysburg, Ohio, at the world's largest photovoltaic manufacturing plant with an estimated capacity of producing enough solar panels each year to generate 100 megawatts of power.

—A family in Morrison, Colorado, installs a 12-kilowatt solar electric system on its home—the largest residential installation in the United States to be registered with the U.S. Department of Energy's http://www.millionsolarroofs.com/ "Million Solar Roofs" program. The system provides most of the electricity for the 6,000—square-foot home and family of eight.

—German Renewable Energy Sources Act came into force and feed in tariff rolled out

2001—NASA's solar-powered aircraft—Helios sets a new world record for non-rocket powered aircraft: 96,863 feet, more than 18 miles high.

—Solar Ark power generation facility was constructed by Sanyo in Japan

2002—Union Pacific Railroad installs 350 blue-signal rail yard lanterns, which incorporate energy saving light-emitting diode (LED) technology with solar cells, at its North Platt, Nebraska, rail yard—the largest rail yard in the United States.

2002—Pathfinder plus—A solar powered aircraft from NASA successfully undertake high altitude telecommunication and aerial imaging.

2003—UNEP launched its solar programme involving banks to promote and install solar home systems in unelectrified home. This was a major boost for the efforts to support distributed solar power.

Thermal Solar energy:

The history of utilization of solar thermal energy in recent times dates back to 1615 when a French engineer Salomon de Caus, built a prototype of solar-driven engine. The invention was a pumping machine, in which the solar energy was used to heat and expand the air to do work of drawing water.

1767—Swiss scientist Horace de Saussure builds the world's first solar collector
1816—Robert Stirling applied for a patent for his economiser at the Chancery in Edinburgh, Scotland. This engine was later used in the dish/Stirling system, a solar thermal electric technology that concentrates the sun's thermal energy in order to produce power
1860—French mathematician August Mouchout proposed an idea for solar-powered steam engines. In the following two decades, he and his assistant, Abel Pifre, constructed the first solar powered engines and used them for a variety of applications. These engines became the predecessors of modern parabolic dish collectors.
1870—Captain John Ericsson developed an improved solar-powered steam engine. Ericsson's initial design had a conical, dish-shaped reflector that concentrated solar radiation onto a boiler and a tracking mechanism that kept the reflector directed toward the sun.
1878—William Adams, an English officer in Bombay, India demonstrated what is known as Power Tower concept. It had flat silvered mirrors arranged in a semicircle with manual tracking of the sun and concentrate the heat to a stationery boiler which powered a 2.5HP steam engine
—William Adams wrote the book titled "Solar Heat—A substitute for Fuel in Tropical Countries"
1882—Abel Pifre, a French engineer ran printing equipment with a steam engine powered by a parabolic solar concentrator
1883—Captain John Ericsson invented a novel method for collecting solar rays-the parabolic trough
1885—Charles Tellier, considered as father of refrigeration, installed flat plate solar collector filled with ammonia on his roof to run a water pump
1891—Clarence Kemp patented the first commercial solar water heater naming it—The Climax
1892—Aubrey Eneas began his solar motor experimentation, formed the first solar power company—The Solar Motor Co.
1897—A third of the homes in Pasadena, California had water heated by the sun.

Period: 1900-1920

In this stage, the research focus of solar thermal energy in the world was on the solar-powered devices in which the thermal energy was utilized with the definite end-use and in higher cost.

- 1904—Henry E. Willsie designed and built for continuous power, two plants, a 6-horsepower facility in St. Louis, Mo., and a 15-horsepower operation in Needles, California.
- 1908—William J. Bailley of the Carnegie Steel Company invented the solar collector with copper coils and an insulated box—similar to its present design.
- 1909—William Bailey patented his solar water heating system that separated the storage tank from an element that collected heat from the sun. This enabled the water to be stored in larger quantities inside the home.
- 1913—Solar Visionary Frank Shuman developed the first solar thermal pumping station in Meadi, Egypt.

Period: 1920-1945

For these 2 decades, the research of solar energy progressed very slowly often grinding to a halt due to the mass utilization of fossil fuels and the Second World War (1939-1945) since the solar energy couldn't satisfy the urgent demand of the energy. Therefore, the research and development of solar energy was to be gradually deserted.

- 1939—Prof Hoyt Hottel built the first active solar house, MIT 1

Period 1945-1965

Solar energy institutes were setup and academic exchanges and exhibitions were held which raised the research upsurge again on solar energy. In this period, great progress was achieved in the research of solar energy.

- 1952—One set of 50kW solar stove was built by French National Research Center
- 1953—Levi Yissar built the first prototype Israeli solar water heater and launched Ner Yah company, Israel's first commercial manufacturer of solar water heating

1953—The oldest solar collector manufacturing company still in business—Solahart was established in Australia.
1955—The foundation theory of selective paints was proposed in the First International Solar Thermal Academic Conference, in which black nickel had been developed as the practical selective paints, contributing to development of high-effective heat collector
1956—The first solar thermal system built in Japan was Unit 1 at the Yanagimachi Solar House
—Architect Frank Bridgers and Donald Paxton designed and built the world's first commercial office building using solar water heating and passive design, the Bridgers-Paxton Building, in Albuquerque New Mexico
1957—Chiryu Heater from Japan produced collectors and is most likely the second oldest collector manufacturing company in the world that is still in business.
1958—The work of Prof Hoyt Hottel and Austin Whillier which derived the solar flat plate collector equation in use today, was presented at the Arizona conference
1960—The worldwide prototype ammonia-water absorbing air conditioning system heated by flat plate heat collector with the capacity of 5 tons was built in Florida of U.S

Period :1965-1973

During this period, the research work on solar energy slowed down the reason being, the technologies of solar energy had entered into the growing stage which was no ripe in process, heavy in investment and lower in effect. Thus it cannot compete with conventional energy, which resulted in the absence of attention and support from the public, enterprise and government.

1969—The Odeillo solar furnace, located in Odeillo, France was constructed. This featured an 8-story parabolic mirror
—There were 4 million solar water heating tanks on the roofs of Japanese homes
1973—The University of Delaware built "Solar One," a PV/thermal hybrid system. In addition to providing electricity, the arrays were like flat-plate thermal collectors; fans blew warm air from over the array to heat storage bins.

1973: Solimpeks—Turkey started the country's first production of solar thermal collector

Period: 1974-1980

After the explosion of Middle East War in 1973, OPEC employed the oil embargo. This crisis made people realize that the existing energy structure should be completely changed and transition to the future energy structure should be speed up.

From that on, many countries, especially the industrialized countries turned their attention towards the support on the research and development of solar energy and other renewable energy technologies. During this period, research and development of solar energy entered into an unprecedented well-developed stage.

1974—The Solar Energy Industries Association (SEIA) formed. The association represents the interests of stakeholders in the solar industries and works in Washington, DC

1976—A group of scientists at the Sydney University teamed by Dr David Mills and Dr Qi-Chu Zang, both senior research fellows in the School of Physics, patented and produced evacuated tube technology for solar heat capturing.

1980—Luz International, an Israeli company founded—the first company to implement the solar thermal power plant technology on a commercial scale

1981—The Fraunhofer Institute for Solar Energy was founded in Germany by Prof Adolf Goetzberger

Period: 1981-1992

As oil prices steadied up, the upsurge on development and utilization of solar energy emerged in 1970s moved towards a stage of low and slows in 1980s. Many countries in the world declined the research budget for solar energy, in particular the U.S. The main reasons resulted in this situation were that the international oil price was corrected in a large range while solar energy product high cost was still remaining as before which was not competitive. With no significant breakthrough on solar energy technologies to increase the efficiency and reduce the cost compounded the erosion of people's confidence to develop and commercially utilize solar energy.

1984—In California the first plant SEGS-1 was built with a parabolic trough solar collector system. The SEGS consist of a solar field with parallel lines of parabolic trough solar collectors connected in series to convert solar energy into heat which warms the oil that flows through the absorber tubes of solar collectors. Hot oil is sent to a heat exchanger which generates superheated steam required to activate a turbo-alternator to produce electricity.

1986—The world's largest solar thermal facility, located in Kramer Junction, California, was commissioned.

1986—Single walled vacuum tubes (evacuated tubes) were developed in Germany by Daimler-Benz, with support from Sunda Solar SEIDO

1992—7.5-kilowatt dish prototype collector system was operational using an advanced stretched-membrane concentrator, through a joint venture of Sandia National Laboratories and Cummins Power Generation.

1992—The international environment and development conference in Brazil by United Nations.

Period: 1993-2005

1994—The first solar dish generator, using a free-piston Stirling engine, was tied to a utility grid.

1998—The world largest solar cooking facility was commissioned Near Mount Abu, Rajasthan India by Brahmakumaris

2003—Global solar annual market crossed 10 GWth

2005—Global total installed capacity of solar thermal cross 100GWth

The early part of the nineties showed resurgence in solar thermal is the US, Germany, Japan and worldwide on a slower pace but gathered momentum during the second half with China entering the fray aggressively. The evacuated tube technology gained more prominence as Chinese companies adopted it in their systems.

Photovoltaic & Thermal: 2005 and beyond

World Bank and its subsidiaries rolled out many successful solar programmes during the early part of this decade in many developing and less developed countries thus giving a good push for solar as the best option for unelectricfied areas and areas with unreliable power supply.

During this period the solar industry saw a high growth all over the world. While the European market consolidated, huge growth plans were set out on China and India. India rolled out the ambitious National solar mission with a target for deployment of 20GW by 2022. China has set itself an ambitious target of 25 to 35% of electricity generation from renewables including 20GW from Solar by 2020. Australia set a target of 20% energy from renewables by 2020. European Union has set a renewable energy target of 20 % by 2020 with southern Europe aiming for a solar power generation installed capacity of about 45GW by 2020. In Europe Germany was leading the way with Spain, Italy, France, UK etc pushing renewables aggressively.

On the PV front companies from China and First Solar from United States established their presence and lead with First solar bringing out thin solar modules capable of directly competing with crystalline modules. China emerging as the largest producer of Solar PV with top rated firms aggressively competing against the west. The commercial crystalline modules surpassed the efficiency of 20% and were available in the market from companies including Sanyo, Sharp etc The Lab efficiency continues to march ahead now crossing the 40% mark.

In the thermal front the sub 70°C water heating market is flourishing with China in the lead, with new applications added and CSP gained a large review all round and raised expectation as a commercially viable business proposition.

CHAPTER 2

SOLAR ENERGY HARNESSING— HOW IT WORKS

> *"I'd put my money on the sun and solar energy. What a source of power! I hope we don't have to wait until oil and coal run out before we tackle that."*
> — *Thomas Edison*

Solar rays descend to earth in abundance. Solar rays contain the solar energy in two forms—(1) Light and (2) Heat. These two basic forms of energy can be converted to useful form and can be a direct substitute for the electricity from the grid. A system which provides solar energy for our household use is known as solar home system (SHS). One question that can come up is how to get the solar energy during night when we use maximum electricity. The answer is simple. We need to store energy for our use. Storage of solar energy is also required to offset the vagaries in supply depending upon the intensity of sunlight. Storing solar energy helps us to use the energy at our convenience.

Hence a solar home system consists of

- Useful(Solar) energy generator /Collector
- Generated energy storage
- Controls for storage and distribution
- Mounting/Safety accessories

Three important things to know when we finalize a solar home system are

(1) How much energy is required by us per day?
(2) How much useful energy will be generated from the system per day?
(3) How much useful energy is stored in the system?
   —Here the term useful energy is important as it depends on the configuration and design of system and its components.

Light to electrical energy:

When sunlight falls on a solar photovoltaic panel it generates electricity in the form of direct current. This is stored in the battery bank. Whenever we need this stored energy it is drawn from the battery through a high efficiency inverter to get AC power useful for our home. This is an indirect form of solar energy usage.

Heat to thermal energy:

When sunlight falls on a thermal collector the heat is directly transferred to the liquid (mostly water) and stored in the hot water tank which is insulated to avoid heat loss. The hot water can be drawn and used when ever required. This is a direct form of solar energy usage.

The previous two paragraphs have given you a basic picture how solar energy can be made useful for our daily use. Broadly building on these two lines more sophisticated solar energy systems are in place now. These even can be used to build large power plants. I will however limit the information here to the ones that are useful for individual home use. In the coming paragraphs detailed description is made for better understanding

Major components of a solar photovoltaic system

Component	What it does
Solar Photovoltaic panel	Collect solar light energy and convert it into dc electricity
Charge controller	Control the charging and discharging of battery
Battery	Draws, stores and deliver energy

Inverter	Converts DC power to AC power
Household lights and appliances	Consume AC power and gives us light and other utility

Solar Photovoltaic panel:
Solar panel consists of an array of solar photovoltaic cells.

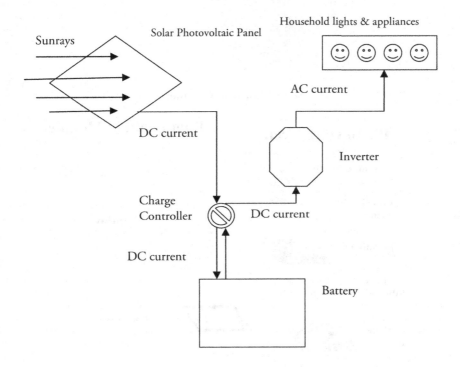

Photovoltaic gets its name from the process of converting light (photons) to electricity (voltage). Photons in sunlight hit the solar cell and are absorbed by semi conducting materials, such as silicon. Electrons (negatively charged) are knocked loose from their atoms, allowing them to flow unidirectional through the material to produce electricity. An array of solar cells converts solar energy into a usable amount of direct current (DC) electricity. In short, the photovoltaic effect occurs when the sunlight energizes electrons in a semiconductor material and causes them to flow through a circuit causing electric current. Solar cells can be broadly classified into monocystalline, polycrystalline and amorphous silicon. The efficiency decreases from monocrystalline to amorphous while ease of manufacturing increases.

Solar Photovoltaic panel

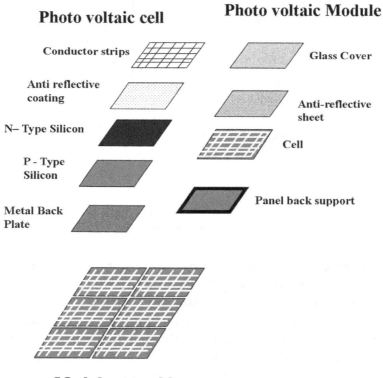

Module assembly

The above left shows the internal components of a photovoltaic cell. The above right shows the internal components of a photovoltaic panel. The lower depicts an array of photovoltaic panels in arrangement.

Charge Controller:

A charge controller limits the rate at which electric current is added to or drawn from batteries. It prevents overcharging and protect against over voltage, which can reduce battery performance or lifespan, and pose a safety risk. It also prevent complete draining ("deep discharging")of a battery, and perform controlled discharges, depending on the battery technology, to protect battery life. Simple charge controllers stop charging a battery when they exceed a set high voltage level, and re-enable charging when battery voltage drops back below that level. Latest advancement in charge controllers includes Pulse width modulation (PWM) and Maximum power point tracker (MPPT) technologies.

Battery:

Batteries accumulate excess energy created by your PV module and store it to be used at night or whenever required. Batteries can discharge rapidly and yield more current than the charging source can produce by itself. Batteries are rated according to their "cycles". Batteries can have shallow cycles between 10% to 15% of the battery's total capacity, or deep cycles up to 50% to 70%. Shallow-cycle batteries, as those for starting a car, are designed to deliver several tens of amperes for a few seconds, then the alternator takes over and the battery is quickly recharged. Deep-cycle batteries on the other hand, deliver a few amperes for tens of hours between charges. These two types of batteries are designed for different applications and should not be interchanged. Deep-cycle batteries are capable of many repeated deep cycles and are best suited for PV power systems. Deep-cycle batteries are specifically designed for energy storage. They have larger and thicker plates. Deep-cycle batteries can withstand discharges having majority of their capacity used before being recharged and survive hundreds of 70% cycles. It is recommended to use 50% as the normal maximum discharge and leave 20% for emergencies. Do not use the bottom 30%. The less deeply you discharge your battery, the longer it will last.

Inverter:

An inverter is an electrical device that converts direct current (DC) to alternating current (AC). The inverter performs the opposite function of a rectifier. A Solar inverter or PV inverter is a type of electrical inverter that is made to change the direct current (DC) electricity from a photovoltaic array/ stored in battery into alternating current (AC) for use with home appliances. It is highly recommended to use pure sine wave inverters for solar home system. This will avoid any hum or overheating of appliances. The efficiency of solar inverters should be high at 90% and above at all load level. The quality of electrical power output that comes to your home depends on this inverter. Hence you should choose the correct inverter intended for renewable energy off grid application.

Other safety components including lightning arrester, bypass circuit breakers, earth leakage relay etc can be used to protect the solar home system against any eventuality.

Household lights & appliances:

Lights and appliances are the load for the PV system. They include the Cfl lights, Tube lights, Ceiling / Pedestal Fan, TV, Refrigerator, Washing machine, Computer etc. It is important to choose the best energy efficient ones to avoid excess energy consumption. The tube lights are available in 50w and 28w which will give the same light out put. Similarly in Fans there are few models which consume 50w and give the same output as a 80w fan. LED lights are an automatic choice in terms of energy efficiency. The cost when compared to cfl may be 4 times. But this is worth due to its longevity and efficiency. When we have a solar powered house it is paramount that we have the best efficient lights and appliances and we run them judiciously. This is limited by the useful energy stored in your system. Always consider the power consumption when you install lights and appliances. These days, there are energy star ratings available for appliances which reflect the efficiency of the appliance.

Major components of a solar thermal water heating system

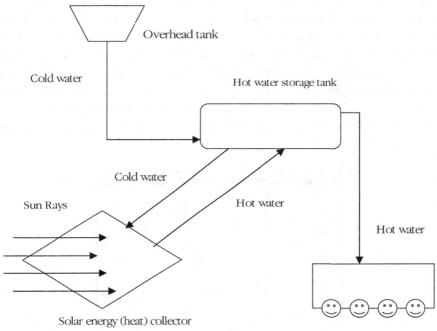

Component	What it does
Solar energy collector	Collects heat energy from the sunrays
Hot water storage tank	Supply cold water to the collector and store the hot water from the collector
Overhead tank	Continuously supply water to the hot water storage tank
Hot water output at home	Consumes solar hot water at our convenience

Solar energy collector: These collectors capture the solar heat energy directly from the sun rays. The heat of the sun rays are transferred to the fluid (mostly water) running in the collector fins/tubes.

Flat Plat collector:

They consist of (1) a dark flat-plate absorber of solar energy (2) a transparent cover that allows solar energy to pass through but reduces heat losses (3) a heat-transport fluid (air, antifreeze or water) to collect heat from the absorber and (4) an insulation backing. The absorber plates of the collector are made up of darkened finned tubes made out of material with good thermal conductivity. These tubes collect the heat energy from the sun rays. The low iron tempered glass provides entry of the sun rays on to the absorber plate. The absorber consists of an absorber fin

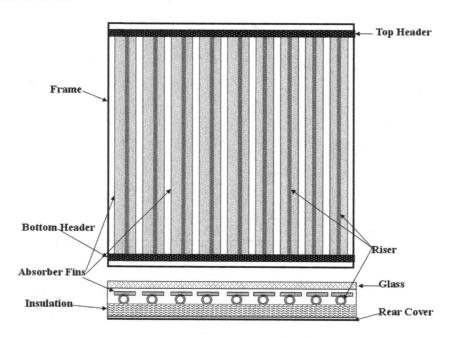

The absorber consists of a thin sheet (Copper, Aluminum or steel to which a black or selective coating is applied) with tubing (riser) attached on the rear. The water (heat transfer fluid) circulates through the absorber tubing to transfer heat from the absorber to the insulated water tank. This may be achieved directly or through a heat exchanger. The cost of these collectors is high at the time of purchase, due to the high content of Copper or Aluminum. For the same reason the resale value / scrap value after 10 to 15 years is also high.

As an alternative to metal collectors, new polymer flat plate collectors are now being produced in Europe. But here in India, these polymer collectors are seldom used. Mostly Copper fins are currently used. Most flat plate collectors have a life expectancy of 15 to 20 years. The flat plate collectors are currently governed by the Bureau of Indian Standards BIS 1233.

Evacuated Tube collector: In an all glass ETC, a selective coated (Aluminum Nitride) glass tube will form the absorber. This will be placed inside an outer tube and both these tubes are fused together with vacuum in between. While the inner tube with selective coating absorbs solar thermal energy, the vacuum in between the tubes restrict the heat loss. Glass-glass tubes have a highly reliable vacuum seal but the two layers of glass reduce the light that reaches the absorber.

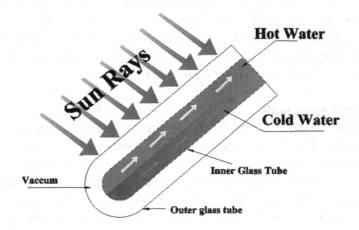

The cost of ETC all glass collectors are only 1/3rd of the flat plate collector.

A combination of Flat Plate and ETC collectors known as heat pipe collectors are also used for various applications. The heat pipe or the Glass-metal tubes allow more light to reach the absorber. This will be housed in a outer glass tube and the space in between evacuated to achieve a high degree of vacuum

The heat from the hot end of the heat pipes is transferred to the transfer fluid (water or an antifreeze mix) which in turn transfer the heat to the colder water in the storage tank.

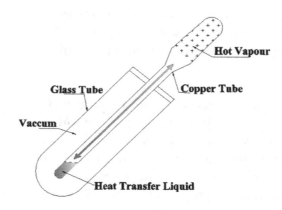

The vacuum that surrounds the outside of the tube greatly reduces convection and conduction heat loss to the outside, therefore achieving greater efficiency than flat-plate collectors, especially in colder conditions. The high temperatures that can occur may require special system design to avoid or mitigate overheating conditions.

Hot water storage tank:

The storage tank enables us to use the hot water at any time during day or night. The main function of this tank is to ensure minimum heat loss from the hot water. This is accomplished by placing 2 tanks concentrically, where the inner tank becomes the hot water holding tank and outer tank the protective casing with adequate insulation in between. Polyurethane foam, rock wool is generally used for insulation. These storage tanks are mounted on rigid structures and grouted.

Both FPC and ETC system employ the principle of thermosyphon for integrated hot water storage tanks

Thermosyphon principle—refers to a method of passive heat exchange based on natural convection . This circulates liquid without the necessity of a mechanical pump. The water become less dense as it gets heated. The denser cold water pushes the less dense hot water to the top there by creating a hot water current towards the top. This principle is used for hot water storage in a solar hot water system.

The tanks are mounted at a higher level than the collectors which enables the thermosyphon to work effectively.

When hot water tanks are not integrated with the collector and located a place of convenience, electric water pumps are used to transfer water from collector to the storage tanks and vice versa.

Overhead tank:

Overhead tank is the source of cold water supply to the solar water heating system. This can be existing overhead tank of the house. The existing overhead tank should not be more than 10 ft higher than the collector, to avoid excessive pressure. Otherwise a small tank needs to be placed at a lower level but above the hot water storage tank to feed cold water to the solar water heating system. The height of the air vent should match with the height of the feeder tank. The over head tank should always have water to continuously feed the solar water heater tank.

Hot water output at home:

Currently the primary usage of solar hot water is for bathing. The bathrooms will have both hot and cold water pipe connection. A mixer valve will facilitate the availability of required temperature for the water for bathing. Since solar heating is not available throughout the year, necessary back up set up should be provided. Electrical heaters that can be fixed inside solar water heater tanks are available or separate electrical geysers can be used. In case of separate geysers, the hot water piping is done separately and connected in parallel to the solar hot water line. Solar hot water can also be used for cooking food in the kitchen, washing clothes etc.

Safety:

One key difference of electrical geyser and solar water heater is that for solar heating the hot water will be stored overnight and used the next day morning. Additionally solar water heater will be located at one place and hot water is distributed to the various usage points in the house. This means that the hot water distribution pipes will be running across the house. The temperature of the hot water in the solar heating system will be around 55 to 70 deg centigrade. Adequate care should be taken for this storage and distribution of hot water. The hot water piping used should have good thermal insulation and should be differentiated from the cold water pipe. Always get the plumbing done by a well trained technician/ plumber.

CHAPTER 3

SOLAR TECHNOLOGIES EXPLAINED

A human being is part of the whole, called by us "Universe" a part limited in time and space. Our task must be to widen our circle of compassion to embrace all living creatures and the whole (of) nature in its beauty"
—Albert Einstein

Solar PV cells can be broadly classified as:

Crystalline cells	Thin Film cells	Emerging PV technologies
1) Monocrystalline Silicon Cells 2) Polycrystalline Silicon Cells	1) Amorphous Silicon Cells 2) Copper indium/ Gallium— Diselenide (CIS or CIGS) Cells 3) Cadmium Telluride (CdTe) Cells	1) Organic Cells 2) Dye Sensitized Cells 3) Multi Junction Concentrators 4) Flexible Solar Nanoantenna

Monocrystalline, or Single Crystal cell:

Single crystal modules are composed of cells cut from a piece of continuous crystal. These are formed out of absolutely pure semi conducting material. The silicon is grown to form a single crystal

The movement of cell efficiency against price is shown below for different types of modules.

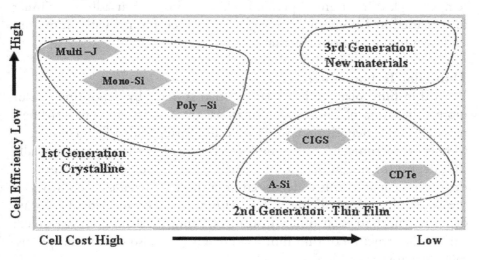

cylinder which is sliced into thin circular wafers. The cells may be fully round or trimmed into other shapes, retaining more or less of the original circle. Because each cell is cut from a single crystal, it has a uniform color which is dark blue. These cells show high efficiency and are the most expensive. The lab efficiency is over 30% while the marketed products have 20%. Very good sunlight is required to get the best output.

Polycrystalline cells:

It is similar with monocrystalline in performance and reliability. Polycrystalline cells are made from similar silicon material by melting, pouring into a mould and recrystallising. This forms a square block that can be cut into square wafers with less waste of space or material than round single-crystal wafers. As the material cools, it crystallizes in an imperfect manner, forming random crystal boundaries. The cells look different from single crystal cells. The surface has a jumbled look with many variations of blue. These cells have lower efficiency and are less expensive than mono crystalline. The lab efficiency is over 20% while the marketed products have 17%. Very good sunlight is required to get the best output.

Twenty-five-year warranties are common for crystalline silicon modules. The construction of finished modules from crystalline silicon cells is generally the same, regardless of the technique of crystal growth. The most common construction is by laminating the cells between a tempered glass front and a plastic backing, using a clear adhesive similar to that used in automotive safety glass. It is then framed with aluminum as shown in the previous chapter.

Thin Film Technologies:

A thin-film solar cell is a solar cell that is made by depositing one or more thin layers of photovoltaic material on a substrate. The thickness range of such a deposition layer is wide and varies from a few nanometers to tens of micrometers. Since the PV cell is made with a microscopically thin deposit of silicon instead of a thick wafer, it would use very little of the precious material. It is deposited on a sheet of metal or glass, without the wasteful work of slicing wafers with a saw. The individual cells are deposited next to each other, instead of being mechanically assembled. This is the structure of thin film technology. (It is also called amorphous, meaning "not crystalline".)

Thin film panels can be made flexible and light weight by using plastic glazing. Some of them perform slightly better than crystalline modules under low light conditions. The lab efficiency is up to 14% while the marketed products have 10%. Hence more space is required to lay compared to crystalline cells. They are also less susceptible to power loss from partial shading of a module. They are cheaper than crystalline cells.

Amorphous silicon

Amorphous silicon is formed by rapidly cooling molten silicon thereby preventing crystal growth and instead forming glass like amorphous silicon. This material lacks internal structure and is flexible, thus can be bend. Bare Amorphous silicon is full of defects, which greatly inhibit the transfer of electric carriers and strongly limit the material's photovoltaic potential. By treating amorphous silicon with hydrogen, these defects are largely corrected. The resulting alloy, referred to as hydrogenated amorphous silicon (a-Si:H), is the basis of most thin-film solar cells seen today.

The single largest disadvantage of amorphous silicon is that the electrical structure of hydrogenated amorphous silicon is metastable, i.e. the material degenerates when exposed to sunlight. This effect, called the Staebler-Wronski effect, causes a-Si:H based solar cells to lose 20% to 30% of their efficiency over the first six months of operation. After this time, the efficiency stabilizes, decreasing only marginally with time.

Copper Indium Gallium Selenide (CIGS):

CIGS is short for copper indium gallium selenide, which are the elements used to make the photoelectric layer in this type of thin film solar cell. CIGS is a compound semiconductor material with great potential for affordable thin-film solar cells. The principle behind the operation of the CIGS solar cell is the same as that of the Silicon solar energy cell. Copper acts to receive electrons, making it the same as the P-type silicon. Selenium provides extra electrons to act in the same way as the N-type silicon.

Although not as efficient as crystalline silicon (the highest recorded efficiency for a CIGS-based cell is 19.9%), CIGS is significantly more efficient than amorphous silicon. It also does not suffer from the degeneration effect that plagues amorphous silicon solar cells. The CIGS layer is usually deposited on a substrate of molybdenum-coated glass, after which it is covered by thin layers of respectively cadmium sulfide (CdS) and zinc oxide (ZnO). These two layers make sure that a proper N-P layer junction exists, allowing the necessary charge separation. Due to the presence of many metals, CIGS cells are susceptible to oxidation; proper encapsulation of the photoactive layers is therefore absolutely necessary to guarantee a sufficiently long lifetime.

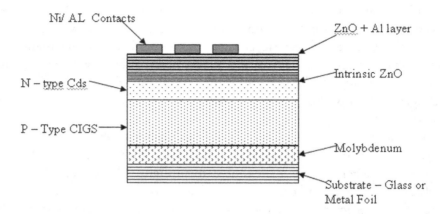

Different layers in a CIGS solar cell

These materials can be placed onto a variety of substrates including thin flexible steel, glass, and various polymers. At present the most widely used substrate is flexible steel as it is the most resistant to the high temperatures needed for the process of laying down the elements onto the backing sheet.

Cadmium telluride (CdTe) cells:

Cadmium telluride (CdTe) cells are formed with the combination of Cadmium, a metallic element and tellurium which is a semi

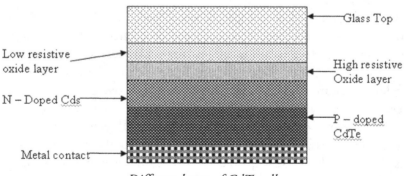

Different layers of CdTe cell

metallic compound. CdTe has an almost perfect band gap for solar energy conversion and can be made by a variety of low cost methods. Safe handling and disposal methods are a must when using these materials due to the toxicity involved. The supply of Telluride seems to be a limiting factor for the production of these cells.

Success of cadmium telluride PV has been due to the low cost achievable with the CdTe technology, made possible by combining adequate efficiency with lower module area costs.

Organic Solar cells:

An organic photovoltaic cell (OPVC) is a photovoltaic cell that uses organic electronics that deals with conductive organic polymers or small organic molecules for light absorption and charge transport.

The plastic itself has low production costs in high volumes. Combined with the flexibility of organic molecules, this makes it potentially lucrative for photovoltaic applications. Molecular engineering can change the gap of energy band, which allows chemical change in these materials. The optical absorption coefficient of organic molecules are high. Hence a large amount of light can be absorbed with a small amount of materials. However these organic photovoltaic cells suffer from lower efficiency, low stability and low strength compared to inorganic photovoltaic cells.

Different types of organic PV cells are listed below.

Single layer organic photovoltaic cell:

Single layer organic photovoltaic cells are the basic form of organic photovoltaic cells. These cells are made by sandwiching a layer of organic electronic materials between two metallic conductors, typically a layer of indium tin oxide (ITO) with high work function and a layer of low work function

| *Electrode 1* *(ITO, Metal)* |
| *Organic electronic material* *(small molecule, polymer)* |
| *Electrode 2* *(Al, Mg, Ca)* |

Single layer organic photovoltaic cell

metal such as Al, Mg and Ca. The basic structure of such a cell is illustrated in the above figure

The difference of work function between the two conductors sets up an electric field in the organic layer. When the organic layer absorbs light, electrons will be excited to Lowest Unoccupied Molecular Orbital (LUMO) and leave holes in the Highest Occupied Molecular Orbital (HOMO) forming excitons. The exciton pairs are separated by the potential created by the different work functions thus pulling electrons to the positive electrode where an electrical conductor used to make contact with a nonmetallic part of a circuit and holes to the negative electrode. The current and voltage resulting from this process can be used to do work.

Bilayer organic photovoltaic cells:

As the name indicates this type of organic photovoltaic cell contains two different layers in between the conductive electrodes in the figure.

These two layers of materials have differences in ionization energy and electron affinity. Therefore electrostatic forces are generated at the interface between the two layers. The materials are chosen properly to make the differences large enough, so these local electric fields are strong, which may break up the excitons much more efficiently than the single layer photovoltaic cells do.

Electrode 1 (ITO, Metal)
Electron Donor
Electron acceptor
Electrode 2 (Al, Mg, Ca)

Multi layer organic photovoltaic cell

The layer with higher electron affinity and ionization potential is the electron acceptor, and the other layer is the electron donor. This structure is also called planar donor-acceptor heterojunctions.

Bulk heterojunction photovoltaic cells:

Electrode 1 (ITO, Metal)
Dispersed Hetrojunction
Electrode 2 (Al, Mg, Ca)

Dispersed junction photovoltaic cell

Here the electron donor and acceptor are mixed together, forming a polymer blend as shown in the figure. When the equivalent length scale condition exists / created, most of the excitons generated in either material may reach the interface, where excitons break efficiently. Electrons move to the acceptor domains then were carried through the device and collected by one electrode, and holes were pulled in the opposite direction and collected at the other side.

Dye sensitized solar cell:

A dye-sensitized solar cell is based on a semiconductor formed between a photo-sensitized anode and an electrolyte, a photo electrochemical system. This cell was invented by Michael Grätzel and Brian O'Regan at the École Polytechnique Fédérale de Lausanne in 1991 and are also known as Grätzel cells. Michael Grätzel won the 2010 Millennium Technology Prize for the invention of the Grätzel cell.

This cell is technically attractive because it is made of low-cost materials and does not require elaborate apparatus to manufacture. Its manufacture could be significantly less expensive than older solid-state cell designs. It can be engineered into flexible sheets and is mechanically robust. Although its conversion efficiency is less than the best thin-film cells, its price/performance ratio (kWh/(m^2·annum·dollar)) is high.

Multi Junction Concentrators:

These cells are developed for high efficiency. These multi junction cells consist of multiple thin films . Each type of semiconductor will have a characteristic band gap energy which causes it to absorb light most efficiently at a certain color, or more precisely, to absorb electromagnetic radiation over a portion of the spectrum. The semiconductors are carefully chosen to absorb nearly the entire solar spectrum, thus generating electricity from as much of the solar energy as possible. The multi junction structure is such that each of several layers capture part of the sunlight passing through the cell. These layers allow the cell to capture more of the solar spectrum and convert it into electricity.

In a single band gap solar cell, efficiency is limited due to the inability to efficiently convert the broad range of energy that photons possess in the solar spectrum. Photons below the band gap of the cell material are lost; they either pass through the cell or are converted to only heat within the material. Energy in the photons above the band gap energy is also lost, since only the energy necessary to generate the hole-electron pair is utilized, and the remaining energy is converted into heat. By utilizing multiple junctions with several band gaps, different portions of the solar spectrum may be converted by each junction at a greater efficiency. Under lab conditions these cells can deliver 35 to 40% efficiency. However lot more need to be done to make this commercially viable

Flexible Solar Nanoantenna:

In August 2008, researchers devised an innovative concept known as a flexible nanoantenna array, which captures infrared radiations emitted by the sun and other energy sources.

A nanoantenna is typically a conducting material that can easily be stamped into a sheet of treated polyethylene. The size of a single nanoantenna is almost 1/25 the diameter of human hair. The nanoantenna has the potential to change the current waste of heat from buildings and electronics. This is because it has the capacity to function as a cooling device while drawing wasted heat from buildings or electronics sans electricity.

The nanotechnology-based solar antenna was developed at the US Department of Energy's Idaho National Laboratory. The scientists developed technologies for imprinting millions of nanoantennas on a large number of flexible materials.

On studying a varied range of materials, such as gold, copper and manganese, researchers demonstrated that nanoanetennas have the capacity to harness up to 92% of energy at infrared wavelengths. Based on this understanding, researchers have created real life prototypes. Through this technology, it is possible to tap energy via invisible mid-infrared radiations, which are soaked and released by the earth during the day. This remedies the main disadvantage of conventional solar cells which fail to utilize solar heat properly and remain idle during the night. Unlike the conventional solar cells, nanoantennas can be used in the dark. This is attributable to the fact that they possess the capacity to capture sunlight and the energy rich infrared radiations of the electromagnetic spectrum, which prevails even after sunset. However, researchers are still working on devising methods for converting the energy captured by these nanoantennas into usable electric energy.

Researchers expect solar nanoantennas to emerge as a more efficient and sustainable alternative to conventional solar panels. It is also estimated that individual nanoantennas will be able to absorb nearly 80% of the energy that is currently available.

Rechargeable Battery—The energy storage device

A battery is a device that converts chemical energy directly to electrical energy and vice versa. This helps in storing electrical energy when available and use the same whenever required.

Batteries can be classified based on

Application	Construction

• Automotive
• Deep cycle
	• Flooded
	• VRLA
	  a) Gel
	  b) AGM

Deep-cycle batteries are the ones ideal for solar energy applications.

Deep cycle batteries:

Deep Cycle Batteries used in Renewable Energy systems are different from car batteries and that difference is critical.

The usage of renewable energy systems by nature is cyclical. Energy is captured and stored, then later consumed, in a regular manner. For example, in a battery-based solar electric system, the energy produced in daylight by the solar panels is stored in the battery bank, which is then used by loads at night or on cloudy days. This repetitive process subjects the batteries to a slow, daily charge and discharge pattern.

Car batteries are not meant to be used in this way. They can release a great deal of their stored energy at once, to start the engine and immediately receive a rapid recharge from the car's alternator. They're not meant to recover their charge slowly, as would happen in a solar electric system.

The depth of discharge for a Car battery is about 20 % while that of deep cycle battery is about 55%. An automotive battery life will come down drastically if used in a renewable energy system. Deep cycle batteries, on the other hand, can be gradually discharged by as much as half of their capacity and can be gradually recharged. When properly maintained, a deep cycle battery can last for four to ten years.

The main two categories of deep cycle batteries are (1) Flooded batteries and (2) Sealed (VRLA) batteries. The flooded batteries need maintenance and sealed batteries are almost maintenance free.

Flooded Lead-Acid (FLA) Deep Cycle Batteries:

Flooded deep cycle lead-acid batteries, also called "wet cells", are commonly used in solar, wind and hydroelectric renewable energy systems. They are often the least expensive type of deep cycle battery and can last the longest. They also come in a wide range of sizes. In FLA batteries, a sulphuric acid solution is the one which reacts with the lead plates in the cells to produce electricity. When FLA deep cycle batteries are recharged electrolysis occurs, producing hydrogen and oxygen gases. These gases escape the cells through the filler/vent caps, which means that the fluid level in the battery goes down. So using FLA batteries in your system should be regularly monitored and maintained by adding distilled water to each cell as needed and ensure that proper ventilation is provided. So it's clear that with FLA batteries, battery maintenance is an essential part of system maintenance.

Sealed (VRLA) lead Acid batteries:

In a VRLA (Valve Regulated Lead Acid) battery the oxygen evolved at the positive plates will largely recombine with the hydrogen ready to evolve on the negative plates, creating water and so preventing water loss. The valve is a safety feature in case the rate of hydrogen evolution becomes dangerously high.

Sealed batteries can be divided into GEL Batteries (contain a semi-solid electrolyte to prevent spillage) and AGM (Absorbed Glass Matt) batteries absorb the electrolyte in a special fiberglass matting. Most valves regulated batteries are under some pressure—1 to 4 psi at sea level.

One advantage of sealed deep cycle batteries is that they can be placed in any orientation: upright, on their sides, and, in some cases, even upside-down. Sealed deep cycle batteries cost more than flooded lead-acid batteries and don't last as many charging cycles; however, they are the preferred choice for applications requiring frequent battery handling, or where the system needs to be left unattended in a remote location. Like FLA batteries, sealed batteries have vents to allow hydrogen to escape when necessary, though this should not normally occur. The difference is that because the batteries are sealed, there is no way to replace the escaped moisture by adding water to the cells; that's why a sealed battery won't last as long as an FLA battery. Gel cell batteries have traditionally been a bit more expensive than AGM batteries; however, they have the unique advantage of performing better in very cold temperatures or in very deep-discharge applications.

In most cases, FLA batteries are an excellent choice for Renewable Energy systems because of their cost-effectiveness, particularly for those systems which require a lot of energy storage. But FLA batteries do require maintenance. Regular monitoring and maintenance will ensure your system functions properly and your batteries live a long, healthy life. On the other hand, if you're using the deep cycle batteries in a remote area where regular maintenance is not possible, sealed batteries will be a better option. Furthermore, sealed batteries conform to situations with space constraints that require you to store your batteries in unusual orientations or where venting is not possible.

Series and Parallel Battery Connections

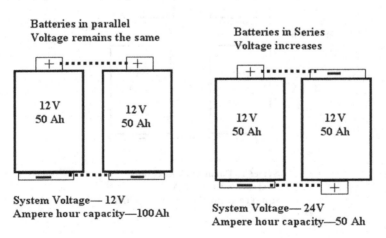

All batteries are sized based on their nominal voltage and Ampere-hour(Ah) of storage.

A combination of Series and Parallel connections increase the battery bank voltage and increase the ampere hour capacity.

For an off-grid system, the capacity of the battery bank must match that of the solar array unless you are using an MPPT charge controller that can accept a higher input voltage than the battery bank voltage. Available sizes in both flooded and sealed batteries vary greatly. The smallest batteries hold less than 20 Ah; the largest approach 2,000 Ah of storage. Usually, these very large batteries are only 2 volts each. If you want to use them in a 48-Volt system, you will need 24 of them, wired in series. Such large batteries could weigh about 100Kg or more. Often size and weight constraints affect the final battery choice.

Useful definitions

Ampere-Hour (Ah): The common unit of measure for a battery's electrical storage capacity, obtained by integrating the discharge

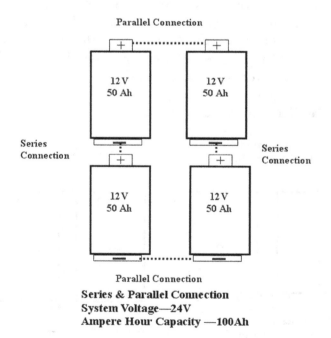

Series & Parallel Connection
System Voltage—24V
Ampere Hour Capacity —100Ah

current in amperes over a specific time period. An ampere-hour is equal to the transfer of one-ampere over one-hour, equal to 3600 coulombs of charge. For example, a battery which delivers 5-amps for 20-hours is said to have delivered 100 ampere-hours.

Capacity: A measure of a battery's ability to store or deliver electrical energy, commonly expressed in units of ampere-hours. Capacity is generally specified at a specific discharge rate, or over a certain time period. The capacity of a battery depends on several design factors including: the quantity of active material, the number, design and physical dimensions of the plates, and the electrolyte specific gravity. Operational factors affecting capacity include: the discharge rate, depth of discharge, cut off voltage, temperature, age and cycle history of the battery. A battery's energy storage capacity can also be expressed in kilowatt-hours (kWh), which can be approximated by multiplying the rated capacity in ampere hours by the nominal battery voltage and dividing the product by 1000. For example, a nominal 12 volt, 100 ampere-hour battery has an energy storage capacity of (12 x 100)/1000 = 1.2 kilowatt-hours.

Cut Off Voltage: The lowest voltage which a battery system is allowed to reach in operation, defining the battery capacity at a specific discharge rate. Manufacturers often rate capacity to a specific cut off, or end of discharge voltage at a defined discharge rate. If the same cut off voltage is specified for different rates, the capacity will generally be higher at the lower discharge rate.

Cycle: Refers to a discharge to a given depth of discharge followed by a complete recharge. A 100 percent depth of discharge cycle provides a measure of the total battery capacity.

Discharge: The process when a battery delivers current, quantified by the discharge current or rate. Discharge of a lead-acid battery involves the conversion of lead, lead dioxide and sulfuric acid to lead sulphate and water.

Charge: The process when a battery receives or accepts current, quantified by the charge current or rate. Charging of a lead-acid battery involves the conversion of lead sulphate and water to lead, lead dioxide and sulfuric acid.

Rate of Charge/Discharge: The rate of charge or discharge of a battery is expressed as a ratio of the nominal battery capacity to the charge or discharge time period in hours. For example, a 4-amp discharge for a nominal 100 ampere-hour battery would be considered a C/20 discharge rate.

Negative (-): Refers to the lower potential point in a dc electrical circuit, the negative battery terminal is the point from which electrons or the current flows during discharge.

Positive (+): Refers to the higher potential point in a dc electrical circuit, the positive battery terminal is the point from which electrons or the current flows during charging.

Open Circuit Voltage: The voltage when a battery is at rest or steady-state, not during charge or discharge. Depending on the battery design, specific gravity and temperature, the open circuit voltage of a fully charged lead-acid cell is typically about 2.1-volts.

Charge controllers:

The primary function of a charge controller in a stand-alone PV system is to maintain the battery at highest possible state of charge while protecting it from overcharge by the PV array and from over discharge by the loads. A charge controller limits the rate at which electrical energy is added to or drawn from electric batteries. These devices may be stand alone or integrated within a battery bank. For solar application charge controllers are also called solar regulators. Although some PV systems can be effectively designed without the use of charge control, any system that has unpredictable loads, user intervention, optimized or undersized battery storage (to minimize initial cost) typically requires a battery charge controller.

Charge regulation: This is the primary function of a battery charge controller, and perhaps the single most important issue related to battery performance and life. The purpose of a charge

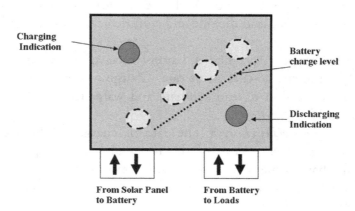

controller is to supply power to the battery in a manner which fully recharges the battery without overcharging. Without charge control, the current from the array will flow into a battery proportional to the irradiance, whether the battery needs charging or not. If the battery is fully charged, unregulated charging will cause the battery voltage to reach exceedingly high levels, causing severe gassing, electrolyte loss, internal heating and accelerated grid corrosion. In most cases if a battery is not protected from overcharge in PV system, premature failure of the battery and loss of load are likely to occur.

A **series charge controller** or **series regulator** disables further current flow into batteries when they are full.

A **shunt charge controller** or **shunt regulator** diverts excess electricity to an auxiliary or "shunt" load, such as an electric water heater, when batteries are full.

Over discharge protection: During periods of below average insolation and/or during periods of excessive electrical load usage, the energy produced by the PV array may not be sufficient enough to keep the battery fully recharged. When a battery is deeply discharged, the reaction in the battery occurs close to the plates, and weakens the bond between the active materials and the plates. When a battery is excessively discharged repeatedly, loss of capacity and life will eventually occur. To protect batteries from over discharge, most charge controllers disconnect the system loads once the battery reaches a low voltage or low state of charge condition.

Simple charge controllers stop charging a battery when they exceed a set high voltage level, and re-enable charging when battery voltage drops back below that level. Pulse width modulation (PWM) and maximum power point tracker (MPPT) technologies are more electronically sophisticated, adjusting charging rates depending on the battery's level, to allow charging closer to its maximum capacity. Charge controllers may also monitor battery temperature to prevent overheating. Some charge controller systems also display data, transmit data to remote displays, and data logging to track electric current flow over time.

Sizing Charge Controllers

Charge controllers should be sized according to the voltages and currents expected during operation of the PV system. The controller must not only be able to handle typical or rated voltages and currents, but must also be sized to handle expected peak or surge conditions from the PV array or required by the electrical loads that may be connected to the controller. It is extremely important that the controller be adequately sized for the intended application. If an undersized controller is used and fails during operation, the costs of service and replacement will be higher than what would have been spent on a controller that was initially right sized for the application.

Under normal conditions a PV module or array produces it's rated maximum power current at 1000 W/m² irradiance and 25 °C module temperature. However the peak array current could be 1.4 times the rated value if reflection conditions exist. Hence, the peak array current ratings for charge controllers should be sized for about 40% more of the nominal peak maximum power current ratings for the modules or array.

The size of a controller is determined by multiplying the peak rated current from an array times this safety factor ie 140%. The total current from an array is given by the number of modules or strings in parallel, multiplied by the module current. In general practice, the short-circuit current (Isc) is generally used instead of the maximum power current (Imp).

Maximum Power Point Tracking Charge controller:

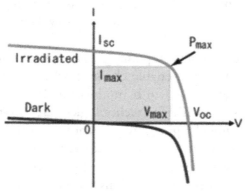

I-V curve for a solar cell, the maximum power point Pmax.

MPPT or **Maximum Power Point Tracking** is an algorithm included in charge controllers used for extracting maximum available power from PV module under certain conditions. The voltage at which PV module can produce maximum power is called 'maximum power point' (or peak power voltage). Maximum power varies with solar radiation, ambient temperature and solar cell temperature. Typical PV module produces power with maximum power voltage of around 17 V when measured at a cell temperature of 25°C, it can drop to around 15 V on a very hot day and it can also rise to 18 V on a very cold day. MPPT checks output of PV module, compares it to battery voltage then fixes what is the best power that PV module can produce to charge the battery and converts it to the best voltage to get maximum current into battery.

MPPT is most effective under these conditions

- Cold weather, cloudy or hazy days: Normally, PV works better at cold temperatures and MPPT is utilized to extract maximum power available from them
- When battery is deeply discharged: MPPT can extract more current and charge the battery if the state of charge in the battery lowers

Dusk to Dawn Charge controllers:

These charge controllers are deployed with systems requiring automatic dusk to dawn operation. It charges the battery during day time and turns on the load at dusk. It turns off the load at dawn. It not only charges the battery from solar panel in the optimum way but also maintains the highest SOC of the battery under charge. These charge controllers are ideal for outdoor lighting applications and remote power solutions.

The principle under which it operates is by monitoring the current generated by Solar panel. The switch over happens when the current comes lower than a set level i.e. the sunlight giving way to darkness and the load (light) is turned on.

While the primary function of a charge controller is to prevent battery overcharge, as seen above many other functions may also be used to get the best from your system. As mentioned earlier we need to bear in mind that a well regulated charging and discharging of batteries is a prime requisite for enhanced battery life and an appropriate charge controller is a must for an uninterrupted operation of the solar home system.

<u>Inverter—AC power generation</u>

An inverter is an electrical device that converts direct current (DC) to alternating current (AC); the converted AC can be at any required voltage and frequency with the use of appropriate transformers, switching, and control circuits. It is so named because early mechanical AC to DC converters were made to work in reverse, and thus were "inverted", to convert DC to AC.

Solid-state inverters have no moving parts and are used in a wide range of applications, from small switching power supplies in computers, to large electric utility high-voltage direct current applications that transport bulk power. Inverters are commonly used to supply AC power from DC sources such as solar panels or batteries.

The inverter performs the opposite function of a rectifier. Inverter are designed to provide 115 VAC/230 VAC from the 12VDC/ 24VDC source.

There are two main types of inverters.

Modified sine wave inverter output waveform is closer to a square wave than to a good sine wave. Due to its simplicity, the cost is low. It is compatible with most electrical devices even though sophisticated electronic devices cannot handle this for extended periods.

Pure sine wave inverter produces a nearly perfect sine wave output (<3% total harmonic distortion) that is essentially the same as utility-supplied grid power. Thus it is compatible with all AC electronic devices. These are high efficient and high in cost.

Basic design

In one simple inverter circuit, DC power is connected to a transformer through the centre tap of the primary winding. A switch is rapidly switched back and forth to allow current to flow back to the DC source following two alternate paths through one end of the primary winding and then the other. The alternation of the direction of current in the primary winding of the transformer produces alternating current (AC) in the secondary circuit.

Output waveforms

The switch in the simple inverter, when coupled to an output transformer, produces a square voltage waveform due to its simple off and on nature as opposed to the sinusoidal waveform that is the usual waveform of an AC power supply. The quality of output waveform that is needed from an inverter depends on the characteristics of the connected load. Some loads need a nearly perfect sine wave voltage supply in order to work properly. Other loads may work quite well with a square wave voltage.

Advanced designs

There are many different power circuit topologies and control strategies used in inverter designs. Different design approaches address various issues that may be more or less important depending on the way that the inverter is intended to be used.

The issue of waveform quality can be addressed in many ways. Capacitors and inductors can be used to filter the waveform. If the design includes a transformer, filtering can be applied to the primary or the secondary side of the transformer or to both sides. Low-pass filters are applied to allow the fundamental component of the waveform to pass to the output while limiting the passage of the harmonic components. If the inverter is designed to provide power at a fixed frequency, a resonant filter can be used. For an adjustable frequency inverter, the filter must be tuned to a frequency that is above the maximum fundamental frequency.

A **Solar inverter** or **PV inverter** is a type of electrical inverter that is made to change the direct current (DC) electricity from a photovoltaic array into alternating current (AC) for use with home appliances and possibly to feed a utility grid.

Solar inverters may be classified into three broad types:

- **Stand-alone inverters**: These are used in isolated systems where the inverter draws its DC energy from batteries charged by photovoltaic arrays and/or other sources, such as wind turbines, hydro turbines, or engine generators. Many stand-alone inverters also incorporate integral battery chargers to replenish the battery. These modified sine wave / pure sine wave inverters provide the AC supply required for the stand alone loads for house or other establishments. The inverter does not interface in any way with the utility grid.
- **Grid tie inverters**: These inverters are always connected to the grid and supply the solar DC converted AC power directly to the grid through net metering. It can simultaneously supply the AC power to the loads at home. These inverters match the phase with a utility-supplied sine wave to ensure flawless

feeding of power to the grid. Grid-tie inverters are designed to shut down automatically upon loss of utility supply, for safety reasons. They do not provide backup power during utility outages. Anti Islanding protection may be provided as a safety measure.

GTIs are often used to convert direct current produced simultaneously by different renewable energy sources, such as solar panels or small wind turbines, into the alternating current used to power homes and business. The technical name for a grid-tie inverter is "grid-interactive inverter". They may also be called synchronous inverters. Grid-interactive inverters typically cannot be used in standalone applications where utility power is not available.

- **Battery backup inverters:** These are special inverters which are designed to draw energy from a battery, manage the battery charge via an onboard charger, and export excess energy to the utility grid. More or less function as a combination of stand alone and grid tie inverter. These inverters are capable of supplying AC energy to selected loads during a utility outage, and are required to have anti-islanding protection.

Anti-islanding protection:

Normally inverters supplying power to the grid will shut off if they do not detect the presence of the utility grid. If, however, there are load circuits in the electrical system that happen to resonate at the frequency of the utility grid, the inverter may be fooled into thinking that the grid is still active even after it had been shut down. This is called islanding. This will result in inverter supplying power to the utility grid even when the grid is down and can be dangerous especially if manual maintenance work is being undertaken. Hence an inverter designed for grid interactive operation will have anti-islanding protection built in; it will inject small pulses that are slightly out of phase with the AC electrical system in order to cancel any stray resonances that may be present when the grid shuts down.

Solar Thermal Technologies

A key component for solar thermal energy utilization is the solar thermal collector. Different types of solar thermal collectors are

Tracking the Sun	Collector type	Absorber type	Concentration ratio	Temp. (° C)
Stationary	Flate Plate collector	Flat	1	35-80
	Evacuated tube Collector	Flat / Tubular	1	50-180
	Compound parabolic Collector	Tubular	1-5	60-240
Single Axis	Linear Fresnel Reflector	Tubular	10-40	50-250
	Parabolic Trough Collector	Tubular	15-45	60-300
	Cylindrical Trough Collector	Tubular	10-50	75-325
Two Axis	Parabolic dish reflector	Point	100-1000	100-500
	Heliostat field collector	Point	150-1500	150-2000
	Dish Stirling Engine	Point	1000-5000	500-800

Flat Plate collectors:

The basic working of Flat Plate Collector is explained in the previous chapter. Flat plate collector as the name indicates, are those collectors that absorb heat energy in a flat plate. Different types of Flat Plate collectors are in use which include

- Glazed type—Use of glazing material (mostly glass) for admitting more radiation on to the absorber
- Unglazed type—A few collectors used for low temperature application employ absorbers without any glazing
- The heat transfer fluids used are water, air, water with antifreeze liquid etc

Evacuated Tube Collectors:

The working principle of evacuated tube collectors is explained in the previous chapter. Three types of evacuated tube collectors are used

- All glass Tube n Tube
- Heat pipe Plate in Glass
- Tube in Glass—U type absorber

The all glass tube captures heat directly to the water and then transfers the hot water to the tank. The heat pipe type uses the liquid—vapour phase change of the heat transfer fluid inside the heat pipe to transfer the heat to the tank.

Compound parabolic Collector:

This is an interesting concept where two sections of parabola face each other which will allow capture high level of heat energy even while Sun's orientation is changing. These concentrators have the capability of reflecting to the absorber all of the incident radiation within wide limits. Compound Parabolic Collectors absorb radiation over a relatively wide range of angles. Radiation entering the aperture within the collector acceptance angle, reaches the absorber through multiple reflections. Two types of absorbers used here are fin type with pipe and tubular absorbers.

Linear Fresnel Collector

LFR technology relies on an array of linear mirror strips which concentrate light on to a fixed receiver mounted on a linear tower. The LFR field is similar to a broken-up parabolic trough reflector, but unlike parabolic troughs, it does not have to be of parabolic shape. Large absorbers can be constructed and the absorber does not have to move. The greatest advantage of this type of system is that it uses flat or elastically curved reflectors which are cheaper compared to parabolic glass reflectors. These are mounted close to the ground, thus minimizing structural requirements.

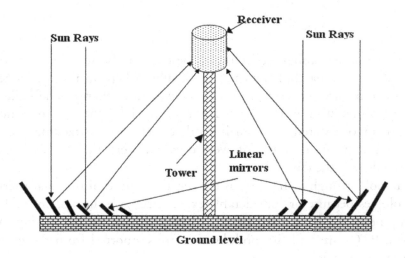

Hardly any moving parts are involved and can reach 200 to 250°C.

The Fresnel collector system will occupy large space for mounting the reflector mirror if their spacing is not optimized with the right height of the receiver. This is a challenge where there is space constraint. Multiple tower systems are other option to improve the collection with the same set of reflector mirrors. Compact linear Fresnel reflector (CLFR) technology is been recently pursued. In this design adjacent linear elements can be interleaved to avoid shading and delivering enhanced power.

Parabolic Trough Collector

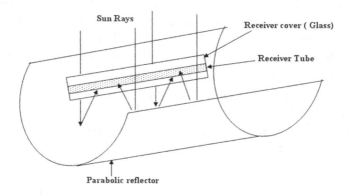

In a parabolic trough collector, as the name indicates, the reflector is in the shape of a parabola and the black tube is kept in the focal line of the parabolic receiver to capture the focused heat energy. The collector can be orientated in an east-west direction, tracking the sun from north to south or orientated in a north-south direction and tracking the sun from east to west. If it is east west oriented, very little adjustment is required during the day.

Parabolic trough technology is the most advanced of the solar thermal technologies because of considerable experience with the systems and the development of a small commercial industry to produce and market these systems. PTCs are built in modules that are supported from the ground by simple pedestals at either end.

Cylindrical Trough Collector

In a cylindrical trough collector, a circular shaped reflector is used to reflect the heat energy to the focus. An single sided / double

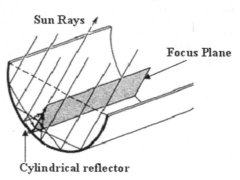

sided absorber sheet is kept on the focus plane where the sunlight will be reflected by the cylindrical trough reflector. The absorber will have the heat transfer mechanism attached. The tracking mechanism provided will track the sun to get the best output.

Parabolic dish reflector (PDR)

A parabolic dish reflector is a point-focus collector that tracks the sun in two axes, concentrating solar energy onto a receiver located at the focal point of the dish. The dish structure must track fully the sun to reflect the beam into the thermal receiver. For this purpose tracking mechanisms are employed by which the collector is tracked in two axes. The receiver absorbs the radiant solar energy, converting it into thermal energy in a circulating fluid. The thermal energy can then either be converted into electricity using an engine-generator coupled directly to the receiver, or it can be transported through pipes to a central power-conversion system.

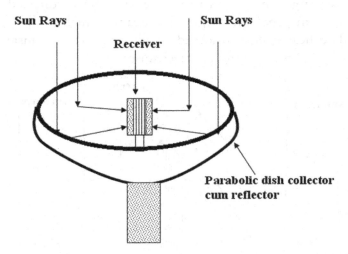

Parabolic-dish systems can achieve temperatures in excess of 1500°C. Because the receivers are distributed throughout a collector field, like parabolic troughs, parabolic dishes are often called distributed-receiver systems.

Parabolic dishes have several important advantages:

1. Because they are always pointing the sun, they are the most efficient of all collector systems;

2. They typically have very high concentration ratio, and thus are highly efficient at thermal-energy absorption and power conversion systems;
3. They have modular collector and receiver units that can either function independently or as part of a larger system of dishes

.Heliostat field collector (HFC)

Heliostat is a device for concentrating the direct solar power on to a receiver. This will deliver very high temperature through focused capturing of solar energy. A highly polished reflector / mirror will reflect the sunlight continuously to a stationary target. The reflective surface will be perpendicular to the bisector of the angle between the direction of the sun and the target—here the central receiver system. This can deliver temperature up to 2000°C. These multiple flat mirrors or heliostats are mounted using altazimuth mounts. This is called the heliostat field or central receiver collector. By using slightly concave mirror segments on the heliostats, large amounts of thermal energy can be directed into the cavity of a steam generator to produce steam at high temperature and pressure. The heat energy absorbed by the receiver is transferred and stored with an appropriate storage mechanism.

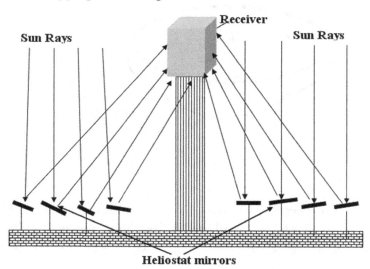

Due to the location and orientation accuracy required for these mirrors, they will be controlled by computerised mechanisms. Large 10 to 20 MW solar thermal power plants can be build using heliostat technology. The central receiver can be coupled with other power generation systems and used as hybrid system. The configuration of the heliostats with the central receiver is critical to get the best output. The selection of the heat transfer fluid, thermal storage medium and power conversion cycle are main characteristics of a central receiver.

Smaller single reflector heliostats are used for day lighting and room warming.

Dish Stirling Engine:
Solar dish Stirling engine use a parabolic dish of mirrors to direct and concentrate the sunlight to a receiver which will act as a power conversion unit using Stirling engine principle. The solar

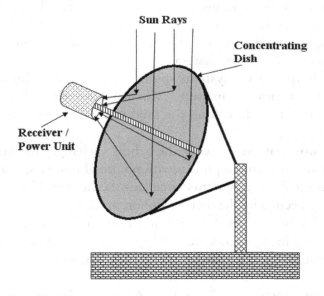

concentrator dish gathers the solar energy coming form the sun.

This concentrated sunlight is reflected on to a thermal receiver where the heat is transferred. This heat is delivered to the engine / generator which operates the stirling engine where the heat of the transfer fluid is converted to piston movement thereby generating mechanical work. This will be converted to electrical energy through a rotating crankshaft or suitable mechanism.

These systems involve two axis tracking and are quite modular. Excellent for configuration with hybrid systems.

Solar water heaters—classification based on heat transfer:

An active system uses an electric pump to circulate the heat-transfer fluid / household water through the collectors

A passive system has no pump but works by the thermo siphon process. The amount of hot water a solar water heater produces depends on the type and size of the system, the amount of sun available at the site, proper installation, and the tilt angle and orientation of the collectors.

An open-loop system (direct) circulates household (potable) water through the collector.

A closed-loop system (indirect) uses a heat-transfer fluid (water or diluted antifreeze, for example) to collect heat and a heat exchanger to transfer the heat to household water.

Open-loop active systems use pumps to circulate household water through the collectors. This design is efficient and lowers operating costs but is not appropriate if your water is hard or acidic because scale and corrosion quickly disable the system.

Closed loop active systems pump heat-transfer fluids (usually a glycol-water antifreeze mixture / water) through collectors. Heat exchangers transfer the heat from the fluid to the household water stored in the tanks.

Closed loop passive systems use a heat exchanger to transfer heat and work by thermo siphon process with no use of pump/motor

Open Loop Passive Systems circulates the household water directly through the collectors by the thermo siphon process

These systems are reliable and relatively inexpensive but require careful planning in new construction because the water tanks are heavy. Proper air vent design is a prerequisite for these systems.

Heat Exchanger system (A closed loop Active/Passive system):

Heat exchangers are deployed in solar water heating systems when the water hardness is very high or when pressurized systems are used. Heat exchanger is a device which uses a secondary fluid to transfer the heat from the collectors to the primary fluid(water for domestic use). This helps in isolating the collectors from the quality / pressure of water to be used.

Here the water coming from the overhead tank or pressure pump will be stored inside the SWH storage tank. The collector collects the heat energy and transfer the heat to the heat exchanger situated inside the storage tank. Any loss of water in the secondary circuit is added through the auxiliary tank.

It can be seen that the collector is isolated from high pressure / poor quality water as this will operate as a secondary circuit and

the storage tank functions as a primary circuit. In active systems a circulation pump is added to the secondary circuit

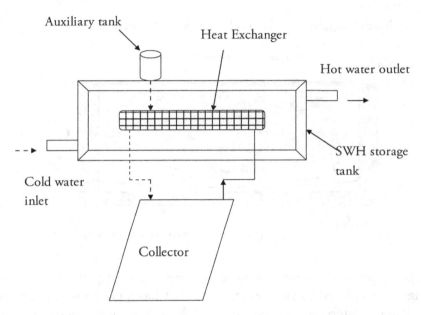

Split solar water heaters (Active systems):

In a split system, as the name indicates the collector and storage tanks are kept separately. The collector—Flat plate type /

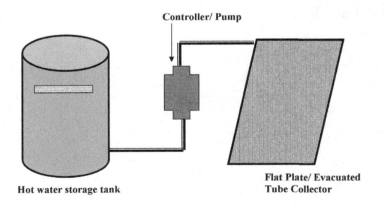

Evacuated tube type—will be kept on the roof mostly as surface mounted. The storage tank will be kept inside the bathroom / wherever space available. The both will be connected by a heat insulated pipes. The water movement to and from the collector to the tank is done with the use of a circulation pump.

Batch Heaters:

Batch heaters (also known as "bread box" or integral collector

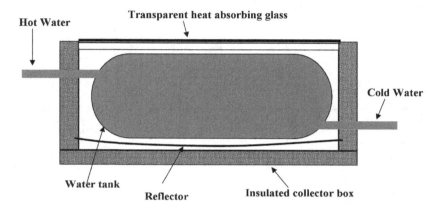

storage systems) are simple passive systems consisting of one or more storage tanks placed in an insulated box that has a glazed side facing the sun. Batch heaters are inexpensive and have few components-in other words, less maintenance and fewer failures.

A batch heater is mounted on the ground or on the roof (make sure your roof structure is strong enough to support it). Some batch heaters use "selective" surfaces on the tank(s). These surfaces absorb sun well but inhibit radiative loss.

CHAPTER 4

SOLAR ENERGY PRODUCTS

"Nature is my god. To me, nature is sacred; trees are my temples and forests are my cathedrals."
—*Mikhail Gorbachev,*
Former President of U.S.S.R

Solar PhotoVoltaic Products:

1) Solar Home System_DC

A solar home system consists of

- Solar Module ranging from 10 Wp to 60 Wp
- Battery from 20 Ah to 80 Ah
- Charge controller 2 Amp to 10 Amp
- Wiring—DC systems require separate wiring
- Connected loads—DC lights, Fan, TV etc

All loads/ equipment run directly from the PV/Battery supply and are designed for 12 V DC operation. The entire system works on DC power. Hence there is minimal loss. DC energy produced by the solar panel is consumed by the DC loads. The limitation being the size of the system and the availability of DC loads/ appliances. Efficient lights and appliances make best use of the limited supply of electricity. Efficient DC fluorescent lights are available in 4 to 16 W power rating, in both tube and compact forms. LED lights are even more efficient, and are now cheap and reliable.

The solar panels are mounted on the roof with charge controller and battery kept inside the house. Wiring is done with thicker cables to interconnect them and to connect them to the loads. Loads located more than 10 meters away from the battery location need to be avoided.

Small DC home systems are used world wide

- In houses without electricity,
- Where power outages are more as a back up,
- At smaller shops,
- In some cases replaces kerosene lamp,
- Schools to rural area
- Administrative offices located remotely

2) Solar Home System_AC

An AC solar home system consists of

- Solar Module ranging from 80 Wp to 1000 Wp or higher
- Battery from 100Ah to 300 Ah
- Inverter with capacity of 350 VA to 4000 VA
- Charge controller 5 Amp to 40 Amp
- Connected loads—AC lights, Fan, TV, Refrigerator, Pump, Mixie etc

The AC system is scalable and can support any amount of load. Constraint will only be the price and space for mounting panels. It is recommended to connect only energy efficient load (lights & appliances) to this PV system. The AC loads are generally less energy efficient and care should be taken while selecting them to power from PV.

For mounting solar panel adequate shadow free space is required on the roof. The solar panels are mounted south facing. To effectively use the solar power, electrical wiring should be done such that the critical loads are separated from the total loads. Good earthing is a must. This should ideally be planned at the time of laying out the cable during the structure construction. Or else there will be need to do a separate wiring when you install the solar home system for AC power.

Solar AC systems are used at

- Residential Homes where the grid power is not connected
- New homes as a power back up during power outages

- Older homes instead of low efficient inverters
- Data centers where continuous power is must
- Hospitals / Clinics in remote area with no access to electricity
- Shops where power outages are rampant
- Green buildings
- Banks and administrative offices
- Apartment complexes for corridor lighting
- Service stations and fuel outlets

3) Solar Wind Hybrid

Solar wind hybrid delivers power throughout the year and hence is an ideal choice of energy harvesting. While solar modules generate good amount of energy during summer, the wind turbine deliver energy during monsoon when the wind is generally high. The resultant hybrid system thus offers an optimal solution at a substantially lower cost compared to an all solar system. The

charge controllers do the balancing act of keeping the battery in good charge position. An inverter, located in the battery bank, changes the current from DC to AC. This electric power is fed to the distribution board.

4) Solar Street Light

Solar street lights have been available for quite a time now, having originally been designed for use in less developed or isolated areas, or perhaps places where the electricity supply has been disrupted by man-made or natural disasters. The solar street light or yard light as it is called consists of a) Solar panel, b) Charge controller c) Battery d) Luminaire e) Pole with arm for mounting the above. The street lights powered by solar energy can be simply and rapidly installed, giving the potential of many years of trustworthy use, with a minimum maintenance.

Solar street lights can deliver exceptional lighting and at the same time, produce considerable economic and environmental savings.

The most recent developments with regard to the technology behind solar street lighting have been in connection with LEDs. Firstly, these consume far less power than the older type of conventional sodium lamp. They have a much longer working life, better color definition and require smaller solar components than sodium lamps / CFL bulbs

The benefits of solar street lights include:

1. Install anywhere there is good daily sunlight for 4 to 5 hrs
2. No digging and laying of cables
3. Automatic switch on and off
4. 365 days of light—no shutdown
5. Low maintenance

Solar street lighting can be used not only on streets, boulevards and highways but can also provide cost-effective, environmentally friendly lighting for both military and civilian security installations; park areas or parking spaces; airports, docks and similar areas or, indeed, for whole communities.

Some of these LED lights feature motion sensors that are triggered once the sun sets by individuals walking in close proximity to the light. If no one is around, the lights conserve energy by remaining off.

Solar street light is one solar product that need very little human intervention for functioning and hence more reliable and long lasting.

5) Solar Garden Light

A solar lamp is a portable light fixture composed of a LED lamp, a photovoltaic solar panel, and one or more rechargeable batteries.

Outdoor lamps are used for lawn and garden decorations. Indoor solar lamps are also used for general illumination (i.e. for garages and the solar panel is detached of the LED lamp). Solar lights used for decoration are sometimes holiday-themed and may come in animal shapes. They are frequently used to mark, footpaths or the areas around swimming pools.

Solar lamps recharge during the day. At dusk, they turn on (usually automatically, although some of them includes a switch for on, off and automatic) and remain illuminated overnight, depending on how much sunlight they receive during the day. Discharging time is generally 8 to 10 hours.

Solar garden lights can be the perfect solution for many outdoor lighting needs. These are commonly purchased and installed in USA and European countries, China being the largest exporter. Solar garden light's installation is as simple as either staking it into the ground or affixing it to a post. There is no need for an electrician, for digging trenches, nor for messing with any wiring.

Solar garden lights utilize the power of the sun to efficiently light outdoor gardens and walkways without any waste of energy.

Garden lighting types include solar lamp lights, fence post lights, yard spot lights, walkway lights, and many more. Nighttime lighting can be a valuable safety tool for visibility of decks, pathways, and stairways. The lights may also simply be used to light up a beautiful fountain or favorite rose bush. The easy installation makes all the possible locations for solar garden lights virtually endless.

6) Solar traffic lights

Solar powered traffic signals are used in many tropical countries like India where electricity supply is not available at places or is erratic. Since traffic signaling systems have to be very reliable, usually a hybrid system using both solar power and grid electricity are used. Solar powered traffic signals are usually designed for 14 hours continuous operation and 10 hours operation in blinking mode (usually when there is less traffic at night). The solar traffic signals have an array of ultra bright long life LEDs. The power consumption of these LED is also much lower.

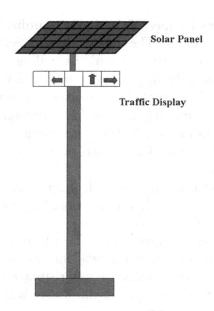

However the solar panel for converting solar energy to electrical energy is usually larger and more efficient. The electrical energy is stored in sealed maintenance free lead acid batteries. The operation of the traffic signal is programmed into a microcontroller/PLC. There will be provision for manual operation also. The signal is visible for a distance of up to 200 meters.

7) Solar water pumping

The system operates on power generated using solar PV (photovoltaic) system. The photovoltaic array converts the solar energy into electricity, which is used for running the motor pump set. The pumping system draws water from the open well, bore well, stream, pond, canal etc.

The system requires a shadow-free area for installation of the solar Panel
Advantages

- No fuel cost—as it uses available free sun light
- No electricity required
- Long operating life
- Highly reliable and durable
- Easy to operate and maintain
- Eco-friendly

Solar water pumps are specially designed to utilize DC electric power from photovoltaic modules. The pumps must work during low light conditions, when power is reduced, without stalling or overheating. Low volume pumps use positive displacement mechanisms which seal water in cavities and force it upward. Lift capacity is maintained even while pumping slowly. These mechanisms include diaphragm, vane and piston pumps. These differ from a conventional centrifugal pump that needs to spin fast to work efficiently. Centrifugal pumps are used where higher volumes are required.

A pump controller is an electronic device used with most solar pumps. It acts like an automatic transmission, helping the pump to start and keeps it from stalling in weak sunlight.

A solar tracker may be used to tilt the PV array as the sun moves across the sky during the day which will increase daily energy gain by about 20% to 55%. Some systems use water storage tanks for simplicity and economy. A float switch added to the system will turn the pump off when the water tank fills, to prevent overflow.

With a diesel generator or when you are on grid, we use a relatively large pump and turn it on only as needed. With solar pumping, we don't have this luxury. Photovoltaic panels are expensive, so we must size our systems carefully. For solar water pumps, first identify how much water you need per day and the vertical lift required. Then the system should be sized and used.

Solar water pumping can be used for:

Domestic Water: Solar pumps are used for private homes, villages, medical clinics, etc. A water pump can be powered by its own PV array, or by a main system that powers lights and appliances. An elevated storage tank may be used, or a second pump called a booster pump can provide necessary water pressure. Or the main battery system can provide storage instead of a tank. Collecting rain water can supplement solar pumping when sunshine is scarce.

Livestock Watering: Cattle ranchers in the Americas, Australia and Southern Africa are enthusiastic users of solar pumps. Their water sources are scattered over vast rangeland where power lines are few and costs of transport and maintenance are high. Some ranchers use solar pumps to distribute water through several miles (over 5 km) of pipelines. Others use portable systems, and move them from one water source to another.

Irrigation: Solar pumps are used on small farms, orchards, vineyards and gardens. It is most economical to power the pump directly from the

PV array (without battery), store water in a tank, and then distribute it by gravity flow. Solar pumps will be most competitive in small installations where combustion engines are least economical.

Solar pumping system should be accompanied by water conservation. Drip irrigation, low water toilets which can reduce total domestic use by half etc should be employed.

8) Solar drip irrigation:

Drip irrigation process provides water directly to plant root systems. Drip irrigation is an example of water-saving irrigation. The increasing human population has rendered water a valuable natural resource with an ever-decreasing supply, so drip irrigation will only become more popular in dry areas. Solar-powered drip irrigation systems use solar powered water pumps and drip irrigation tubing and emitters to provide water sources in difficult climates. Solar-powered drip irrigation systems pump water from deep within the earth and up to basins attached to drip irrigation tubing. This method helps deliver water to crops year-round even in areas of the world that experience long dry periods. Solar-powered drip irrigation produced benefits even in harsh drought-stricken African climates. Many number of solar powered drip irrigations systems are installed all over the world.

9) Solar Lantern

Solar lantern is an ideal replacement for the grand old kerosene lamp. The battery, the charging—discharging circuit and the bulb

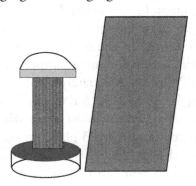

is housed in lantern housing and the module is separately provided. The module can be located at a convenient place and the lantern can be charged during day time.

The greatest advantage of the solar lantern is its portability. Lakhs of solar lanterns are being sold every year and millions are in use throughout the world. Lanterns with LED bulbs are available recently and are more reliable. The battery in Solar lantern are mostly Sealed Lead Acid type and care should be taken to keep it adequately charged—should not keep fully discharged for more than 48 hrs. Currently solar lanterns are mostly used in rural areas.

10) Solar Plug & play systems

The availability of plug-and-play solar kits make limited-use applications both easy and affordable. They do not require the user

to learn the electrical and mechanical ins and outs of solar energy. These consists of a solar panel (collector), a battery and control housing (power pack)and load (lights, fan and accessories).The unit's solar panel is placed in a suitable location to draw energy from the sun's rays. At night the portable power pack is moved in the house for use in a single room for lighting or as a charging station for small laptop computers, cell phones, and similar electronics. With an adequately sized system you will be able to power your reading lamp, fan etc for the evening without worrying about power outages, and the system most likely will have enough power left over to charge your cell phone, MP3 player or similar small devices.

Selecting a plug-and-play solar system means that you have a portable, self-contained unit you can take with you for remote use, like a day at the beach or a weekend at the cabin. These units can help you undertake your electrical tasks "off grid," while giving you a reliable, portable source of clean, renewable energy.

11) Solar DC Refrigerators:

As any solar appliance a PV DC Solar Fridge requires three crucial components for it to run. Firstly, solar panels draw power of sun and send out DC current to the batteries. The batteries are regulated by solar charge controllers that make sure that proper power goes to it and there is no overcharging. So, typically a DC Solar Fridge works using a combination of solar panels and lead batteries to store energy for overcast conditions and at night.

The fridge itself comprises of two major parts—the compressor circuit and body. Solar powered fridges are very energy efficient and not too noisy. The compressor is optimized for cooling to maximum with least power consumption. The other critical component of the fridge is the insulated body. A solar fridge body normally has a polyurethane insulation and coated steel cabinets. The body of the fridge should have a good insulation to limit the compartment heat gain to absolute zero. Further the gasket and flange are also made for ensuring that no cooling loss occurs when the doors are closed.

A solar powered DC fridge may have a cooling fan to dissipate the low temperature at the evaporator coils and maintain uniform cooling inside the fridge. Due to its over design these refrigerators cost up to USD 1000 for a 135 liter ones.

12) Solar fencing

The solar fencing is a metal fence which gets energized with a

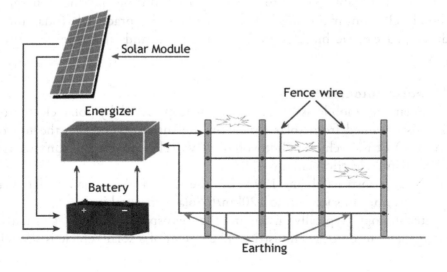

energizer. The energizer is powered by a battery which gets charged by an array of solar panel. The energizer will give out a limited level of shock when any unwanted elements touch the fence. Proper earthing is done to avoid any excessive damage.

Solar fencing are used for

- High security fencing
- Security fencing
- Animal fencing
- Agriculture fencing

13) Solar bicycles and motorcycles

A solar bicycle or tricycle has the advantage of very low weight and can use the rider's foot power to supplement the power generated by the solar panel roof. This way, a simple and inexpensive vehicle can be driven without the use of any fossil fuels.

Solar photovoltaic helped power India's first Quadricycle developed since 1996 in Gujarat state's SURAT city.

The first solar "cars" were actually tricycles or quadricycles built with bicycle technology. These were called solarmobiles at the first solar race, the Tour de Sol in Switzerland in 1985 with 72 participants, half using exclusively solar power and half solar-human-powered hybrids. A few true solar bicycles were built, either with a large solar roof, a small rear panel, or a trailer with a solar panel. More practical solar bicycles were built with foldable panels to be set up only during parking. Charging the bicycle from mains powered by solar was also practiced. Today more developed electric bicycles are being tried out and these use very little power.

14) Solar Automobiles

Solar automobile is an electric vehicle powered by solar electricity. This is obtained from solar panels on the surface (generally, the top or window) of the vehicle. Photovoltaic (PV) cells convert the sun's energy directly into electrical energy

Ned, constructed in 1999 by the South Australian Solar Car Consortium, can speed up to 120km/h.Solar cars combine
technology typically used in the aerospace, bicycle, alternative energy and automotive industries. The design of a solar vehicle is severely

limited by the amount of energy input into the car. Most solar cars now have been built for the purpose of solar car races.

. The two most notable solar car races are the World Solar Challenge and the North American Solar Challenge, road rally-style competitions contested by a variety of university and corporate teams.

The World Solar Challenge features a field of competitors from around the world who race to cross the Australian continent, over a distance of 3000 km. Speeds of the vehicles have steadily increased. The high speeds of 2005 race participants led to the rules being changed for solar cars starting in the 2007 race.

The North American Solar Challenge, previously known as the 'American Solar Challenge' and 'Sunrayce USA', features mostly collegiate teams racing in timed intervals in the United States and Canada. This race also changed rules for the most recent race due to teams reaching the regulated speed limits.

The Dell-Winston School Solar Car Challenge is an annual solar-powered car race for high school students. The event attracts teams from around the world, but mostly from American high schools. The race was first held in 1995. Each event is the end product of a two-year education cycle launched by the Winston Solar Car Team. In odd-numbered years, the race is a road course that starts at the Dell Diamond in Round Rock, Texas; the end of the course varies from year to year. In even-numbered years, the race is a track race around the Texas Motor Speedway. Dell has sponsored the event since 2002.

The South African Solar Challenge is an epic, bi-annual, two-week race of solar-powered cars through the length and breadth of South Africa. Teams will have to build their own cars, design their own engineering systems and race those same machines through the most demanding terrain that solar cars have ever seen. In 2008 the event was endorsed by International Solarcar Federation (ISF), Fédération Internationale de l'Automobile (FIA), World Wildlife Fund (WWF) making it the first Solar Race to receive endorsement from these 3 organizations.

There are other distance races, such as Suzuka, Phaethon, and the World Solar Rally. Suzuka is a yearly track race in Japan and Phaethon was part of the Cultural Olympiad in Greece right before the 2004 Olympics.

15) Solar Boat :

Planet Solar, a 31 meter long catamaran, was unveiled in Kiel, Germany, and its deck just so happens to be completely covered with photovoltaic panels, making it the world's largest solar powered boat. The makers of the boat say, "Planet Solar wants to show that we can change, that solutions exist and that it isn't too late. Future generations are looking to us; our choices will mark the future of humanity."

Built at the Knierim Yacht Club in Kiel in northern Germany, the Planet Solar is a 31 by 15 meter catamaran that can expand to 35 by 23 meters when the flaps at the stern and the sides are extended.

Copyright: © PlanetSolar

The deck is completely covered in 500 sq. meters of solar panels with the cockpit sticking out from the top; they will be rolling into port without emissions. Manned by two crew members, the catamaran can accommodate up to 50 people on their world voyage. The makers are expecting the boat to get a top speed of 15 knots and an average of 8 knots.

16) Solar Aircraft

Flying with the help of direct sun was passion and dream of many and several experimental flights were undertaken since the eighties. A more serious attempt is Solar Impulse, a European long-range solar powered plane project initiated in 1999 being undertaken at the École Polytechnique Fédérale de Lausanne. The project is co-promoted by Bertrand Piccard, who co-piloted the first balloon to circle the world non-stop. This project hopes to repeat that feat with a piloted fixed-wing aircraft using only solar power. The first aircraft is Solar Impulse HB-SIA. This single seated aircraft weighs only 1.8 tons, and features a 208—foot wingspan, the same as an Airbus A340. The technical datasheet

Copyright: © Solar Impulse/ Revillard/ rezo.ch

talks about an average speed of 37.8 knots (70 km/h), and a maximum altitude of 8,500 meters above sea level. The night flight powered by solar energy, captured and stored during the day and used to run four electric motors; the plane's wings have about 12,000 solar cells. The aircraft was unveiled in June 2009 and in July 2010 it created history as the first night flight in the history of solar aviation in a 26 hour flight.

This astonishing aircraft flew across to Switzerland, Brussels and then to Morocco.

Building on the experience of this prototype, a slightly larger follow-on design (HB-SIB) is planned for 2015 to circumnavigate of the globe in 20-25 days.

Solar Thermal Products

1) Domestic solar water heaters

A solar water heater consists of a collector to collect solar energy and an insulated storage tank to store hot water. The concepts and the product have been explained in the previous chapters. The most common type used are Flat Plate Collector type (FPC) and Evacuated Tube Collector type (ETC).

Salient Features of Solar Water Heating System

Solar Hot Water System turns cold water into hot water with the help of sun's rays.

- Around 60 deg.—80 deg. C temperature can be attained depending on solar radiation, weather conditions and solar collector system efficiency
- Provides Hot water for homes, hostels, hotels, hospitals, restaurants, dairies, industries etc.
- Can be installed on roof-tops, building terrace and open ground where there is no shading, south orientation of collectors is possible and over-head tank is above SWH system
- SWH system generates hot water on clear sunny days (maximum), partially clouded (moderate) but not in rainy or heavy overcast day
- Only soft and potable water can be used
- Stainless Steel is used for small tanks whereas Mild Steel tanks with anticorrosion coating inside are used for large tanks
- Solar water heaters (SWHs) of 100-300 liters capacity are suited for domestic application.
- Larger systems can be used in restaurants, guest houses, hotels, hospitals, industries etc.

Fuel Savings:

A 100 liters capacity SWH can replace an electric geyser for residential use and saves 1500 units of electricity annually.

Avoided utility cost on generation:

The use of 1000 SWHs of 100 litres capacity each can contribute to a peak load saving of 1 MW.

Environmental benefits:
A SWH of 100 litres capacity can prevent emission of 1.5 tons of carbon dioxide per year.

Approximate cost: Rs.15000-20,000 for a 100 litres capacity system and Rs.110-150 per installed litre for higher capacity systems

Payback period: 3-4 years when electricity is replaced

Though the initial investment for a solar water heater is high compared to available conventional electric water heaters, the return on investment has become increasingly attractive with the increase in prices of conventional energy. The pay back period depends on the site of installation, utilization pattern and fuel replaced.

2) Commercial solar water heaters

The commercial solar water heating systems are larger ones delivering normally 1000 to 5000 LPD systems and their multiples. These are used for big hotels, hospitals, hostels etc. The advantage for hotels is that the hot water can be used for bathing as well as for washing plates and utensils. The solar water heating system will deliver a temperature of 60 to 75 deg depending on the sunlight and hot water consumption. The hot water piping

holds the key for best performance of the system. Adequate pumps, temperature controls are to be adopted to maximize the usage. It is often found that accessories are compromised which result in customer dissatisfaction even though sufficient hot water is produced by the collector and transferred to the storage tank. Hence it is advisable to have all the hot water distribution piping and accessories are listed and finalized at the time of purchasing the solar hot water system. Results have shown that it is better always to have a higher capacity system installed to offset the changes in hot water generation and customer usage pattern.

3) Industrial water heating

The potential applications of SWH in the industries are:

- Pre-heating of boiler feed water: In this application, either full or part of the boiler feed water is heated in solar water heaters to a temperature of 60-80°C before being supplied to the boiler. This replaces part of the fuel used in the boiler. From the point of integration with the existing process, this is simple to implement. It is highly economical to use solar to replace petroleum fuels. The pay-back period after considering depreciation and subsidy benefits for a furnace oil based industry is reported to be around 3 years, while for a coal using industry, the pay-back period is around 5-6 years.
- Heating of process hot water: There are several industrial processes e.g. electroplating, textile dyeing, cleaning/ degreasing, drying, etc. which require hot water below 100°C. In these applications, solar water heater in conjunction with a conventional water heating system can be used. In such cases, the fuel savings are much larger, but the system integration generally is more complex. The economics of this option becomes favourable when petroleum fuel or electricity is being used.
- Canteen applications: Industrial canteens exist in most of the organized small, medium and large industries. Hot water is required for both cooking as well as washing of utensils, hands etc. Solar water heating is an ideal option.

4) Solar cookers

A solar oven or solar cooker is a device which uses sunlight as its energy source. Solar Cookers are a form of outdoor cooking and are often used in situations where minimal fuel consumption is important, or the danger of accidental fires is high. There are a variety of types of solar cookers: over 65 major designs and hundreds of variations of them. The basic principles of all solar cookers are:

- Concentrating sunlight: Some device, usually a mirror or some type of reflective metal, is used to concentrate light and heat from the sun into a small cooking area, making the energy more concentrated and therefore more potent.
- Converting light to heat: Any black on the inside of a solar cooker, as well as certain materials for pots, will improve the effectiveness of turning light into heat. A black pan will absorb almost all of the sun's light and turn it into heat, substantially improving the effectiveness of the cooker. Also, the better a pan conducts heat, the faster the oven will work.
- Trapping heat: Isolating the air inside the cooker from the air outside the cooker makes an important difference. Using a clear solid, like a plastic bag or a glass cover, will allow light to enter, but once the light is absorbed and converted to heat, a plastic bag or glass cover will trap the heat inside. This makes it possible to reach similar temperatures on cold and windy days as on hot days.

Most solar cookers use two or all three of these principles in combination to get temperatures sufficient for cooking. The top can usually be removed to allow dark pots containing food to be placed inside. One or more reflectors of shiny metal or foil-lined material may be positioned to bounce extra light into the interior of the oven chamber. Cooking containers and the inside bottom of the cooker should be dark-colored or black. Inside walls should be reflective to reduce radiative heat loss and bounce the light towards the pots and the dark bottom, which is in contact with the pots.

Solar Box Cooker

Solar Box Cooker resembles a box type cooking apparatus. Invented by Horace de Saussure, a Swiss naturalist, as early as 1767, solar box cookers gained popularity since the 1970s. The top is transparent and the sides are insulated with crumpled newspapers, wool, rags, dry grass, sheets of cardboard, etc. The transparent top is either glass, which is durable but hard to work with, or an oven cooking bag, which is lighter, cheaper, and easier to work with, but less durable. The solar box cooker typically reaches a temperature of 150 °C. This is not as hot as a standard oven, but still hot enough to cook food over a longer period of time. It is best to start cooking before noon, though. Depending on the latitude and weather, food can be cooked either early or later in the day. The cooker can be used to warm food and drinks and can also be used to pasteurize water or milk. If you use an indoor stove for your actual cooking, you can save significant fuel by using the solar cooker to preheat the water to be used for cooking grains, soups, etc., to nearly boiling.

Panel cookers

Panel solar cookers are very inexpensive solar cookers that use shiny panels to direct sunlight to a cooking pot that is enclosed in a clear plastic bag. A common model is the CooKit. Developed in 1994 by Solar Cookers International, it is often produced locally by pasting a reflective material, such as aluminum foil, onto a cut and folded backing, usually corrugated cardboard. It is lightweight and folds for storage. The CooKit is considered a low-to-moderate temperature solar cooker, easily reaching temperatures high enough to pasteurize water or cook grains such as rice. On a sunny day, one CooKit can collect enough solar energy to cook rice, meat or vegetables to feed a family with up to three or four children. Larger families use two or more cookers. To use a panel cooker, it is folded into a bowl shape. Food is placed in a dark-colored pot, covered with a tightly fitted lid. The pot is placed in a clear high temperature plastic bag and tied, clipped, or folded shut.

The panel cooker is placed in direct sunlight until the food is cooked, which usually requires several hours for a full family-sized meal. For faster cooking, the pot can be raised on sticks or wires to allow the heated air to circulate underneath it.

A recent development is the Hotpot developed by US NGO Solar Household Energy, Inc. . The cooking vessel in this cooker is a large clear pot with a clear lid into which a dark pot is suspended. This design has the advantage of very even heating since the sun is able to shine onto the sides and the bottom of the pot during cooking. An added advantage is that the clear lid allows the food to be observed while it is cooking without removing the lid. The Hotpot provides an alternative to using plastic bags in a panel cooker.

Solar kettles

Solar kettles are solar thermal devices that can heat water to boiling point through the reliance on solar energy alone. Primarily consists of a water holding jar or collector and a reflector. Some of the recently developed kettle use evacuated solar glass tube technology to capture, accumulate and store solar energy needed to power the kettle. Since solar vacuum glass tubes work on accumulated rather than concentrated solar thermal energy, solar kettles only need diffused sunlight to work and needs no sun tracking at all.

Cookers with parabolic reflectors:

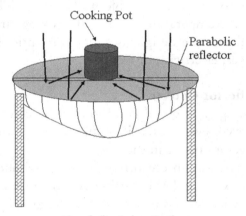

Parabolic Solar Cooker

Parabolic cookers reach high temperatures and cook quickly, but require frequent adjustment and supervision for safe operation. Several hundred thousand exist, mainly in China. They are especially useful for large-scale institutional cooking.

Parabolic reflectors that have their centers of mass coincident with their focal points are useful. They can be easily turned, to follow the sun's motions in the sky, rotating about an axis that passes through the focus. The cooking pot therefore stays stationary.

Cookers with spherical reflectors

The Solar Bowl is a unique concentrating technology used by the Solar Kitchen in Auroville, India. Unlike nearly all concentrating technologies that use tracking reflector systems, the solar bowl uses a stationary spherical reflector. This reflector focuses light along a line perpendicular to the sphere's surface and a computer control system moves the receiver to intersect this line. Steam is produced in the solar bowl's receiver at temperatures reaching 150 °C and then used for process heat in the kitchen where 2,000 meals are prepared daily.

Hybrid cookers

A hybrid solar oven is a solar box cooker equipped with a conventional electrical heating element for cloudy days or nighttime cooking. Hybrid solar ovens are therefore more independent. However, they lack the cost advantages of some other types of solar cookers.

A hybrid solar grill consists of an adjustable parabolic reflector suspended in a tripod with a movable grill surface. These outperform solar box cookers in temperature range and cooking times. When solar energy is not available, the design uses any conventional fuel as a heat source, including gas, electricity, or wood.

The Scheffler-Reflector Cooker

To make cooking simple and comfortable the cooking-place should not have to be moved, even better: it should be inside the house and the concentrating reflector outside in the sun.

The best solution was an eccentric, flexible parabolic reflector which rotates around an axis parallel to earth-axis, synchronous with the sun. Additionally the reflector is adjusted to the seasons by flexing it in a simple way.

The reflector is a small lateral section of a much larger paraboloid. The inclined cut produces the typical elliptical shape of the Scheffler-Reflector. The sunlight that falls onto this section of the paraboloid is reflected sideways to the focus located at some distance of the reflector.

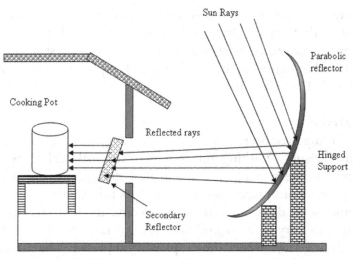

Scheffler-Reflector Cooker for community Kitchen

In the course of the seasons the incident angle of the solar radiation varies + / -23, 5° in relation with the perpendicular to earth-axis. The paraboloid has to perform the same change of inclination in order to stay directed at the sun.

Adjustment of the reflector-shape has to be done manually every 2-3 days. When all concentrated light enters the opening of the cooking-place installed at the focal point the correct reflector position is achieved.

After passing the opening the light is redirected by a small reflector (secondary reflector) to the black bottom of the cooking pot. There it is absorbed and transformed into heat. The efficiency for cooking, i.e. heating water from 25°C to 100°C, can reach up to 57% and depends on the cleanliness of the reflector-surface and the state of insulation of the cooking-pot. The cooking power varies with the season. There are many options for the design of the cooking-place. Mostly it is integrated into a kitchen building and provides the possibility to use firewood for cooking when the sun doesn't shine. Depending on the type of food which is cooked there is no need for a secondary reflector. This increases the efficiency and simplifies maintenance. Instead of a cooking-place a backing-oven, steam-generator or heat-storage can be installed at the focal-point.

The first well functioning Scheffler-Reflector (size: 1,1m x 1,5m) was built by Wolfgang Scheffler in 1986 at a mission-station in North-Kenya and is still in use.

The biggest installation of this kind is used in India at a yoga-centre Abu Road, Rajasthan by Brahma Kumaries to cater for about 18 000 persons. Here the steam is also used as a medium of storage (2 hours full power without sunshine). The Tirupati Temple in Andhra Pradesh is equipped with 105 reflectors.

5) Solar Dryer:
Solar drying has been practiced since long time. This was natural drying in the sun. Working on the solar thermal technologies many inventions have been made for accelerated drying. Drying of crops is being done at various places and the awareness is increasing.

A simple solar dries consists of a collector, duct to take heat to the drying zone, the drying zone and an exhaust manifold. The drying zone will have drying trays where the crops will be placed. The air inside the heat collector box will get heated and move up. The cold air outside will move in. The hot air from the collector will move to the drying zone and heat up the zone resulting in the crops getting heated. In some cases forced circulation is also done to get better result. Solar dryers are used in agriculture for food and crop drying, for industrial drying process etc. The dryers not only save energy but also save lot of time, occupying less area, improve quality of the product, make the process more efficient and protect environment. Solar drying can be used for the entire drying process or for supplementing artificial drying systems, thus reducing the total amount of fuel energy required.

Solar dryer is a very useful device for

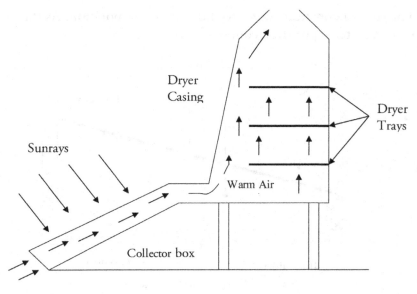

- Agriculture crop drying
- Food processing industries for dehydration of fruits, potatoes, onions and other vegetables,
- Dairy industries for production of milk powder, casein etc.
- Seasoning of wood and timber.
- Textile industries for drying of textile materials.

6) Solar still

Solar distillation is a tried and tested technology. The first known use of stills dates back to 1551 when it was used by Arab alchemists. The first "conventional" solar still plant was built in 1872 by an engineer named Charles Wilson in the mining community of Las Salinas in northern Chile. This still was a large basin-type still used for supplying fresh water using brackish feed water to a nitrate mining community. The plant used wooden bays which had blackened bottoms using logwood dye and alum. The total area of the distillation plant was 4,700 square meters. On a typical summer day this plant produced 5 liters of distilled water per square meter of still surface, or more than 23,000 liters per day. This first still's plant was in operation for 40 years!

The sun's energy heats water to the point of evaporation. As the water evaporates, water vapor rises, condensing on the glass

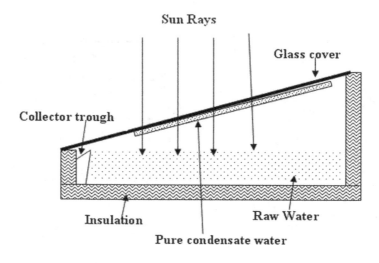

surface for collection. This process removes impurities such as salts and heavy metals as well as eliminates microbiological organisms. The end result is water cleaner than the purest rainwater. Purified drinking water is collected from the output collection port.

The still will continue to produce distillate after sunset until the water temperature cools down. Feed water should be added each day that roughly exceeds the distillate production to provide proper flushing of the basin water and to clean out excess salts left behind during the evaporation process. The intensity of solar energy falling on the still is the single most important parameter affecting production. Solar stills use natural evaporation and condensation, which is the rainwater process. Solar distillers are used to make drinking water from seawater.

Typical efficiencies for single basin solar stills approach 60 percent. General operation is simple and requires facing the still towards solar noon, putting water in the still every morning to fill and flush the basin, and recovering distillate from the collection reservoir (for example, glass bottles). Stills are modular and for greater water production requirements, several stills can be connected together in series and parallel as desired.

7) Solar Air conditioning:

Absorption refrigeration cycle is most commonly used to cool space by solar thermal systems.

Absorption refrigeration uses a heat source to provide the energy needed to drive the cooling system. Absorption refrigerators are a popular alternative to regular compressor refrigerators where

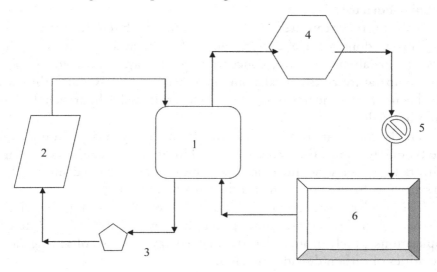

1—Absorption Heat exchanger system
2—Solar thermal collectors
3—Circulation Pump
4—Condenser
5—Expansion Valve
6—Evaporator: Takes heat from the surroundings and create cooling effect

electricity is unreliable, costly, or unavailable, where noise from the compressor is problematic, or where surplus heat is available (e.g., from turbine exhausts or industrial processes). For solar absorption systems the heat source is Solar thermal collectors—FPC, Heat Pipe etc.

Both absorption and compressor refrigerators use a refrigerant with a very low boiling point (less than −18 °C). In both types, when this refrigerant evaporates (boils), it takes some heat away with it, providing the cooling effect. The main difference between the two types is the way the refrigerant is changed from a gas back into a liquid so that the cycle can repeat. An absorption refrigerator changes the gas back into a liquid using a method that needs only heat, and has no moving parts. A compressor refrigerator uses an electrically-powered compressor with moving parts for this purpose. The other difference between the two types is the refrigerant used. Compressor refrigerators typically use an HCFC or HFC, while absorption refrigerators typically use ammonia or more recently Lithium Bromide solution.

8) Solar Furnace

A solar furnace captures sunlight to produce high temperatures.

This is done with a curved mirror (or an array of mirrors) that acts as a parabolic reflector, concentrating sun rays onto a focal point. The temperature at the focal point may reach 3,500 °C, and this heat can be used to generate electricity, melt steel, make hydrogen fuel or nanomaterials.

The largest solar furnace is at Odeillo in the Pyrenees mountains in the Basque region of France, opened in1970. It employs an array of plane mirrors to gather sunlight, reflecting it onto a larger curved mirror. The rays are then focused onto an area the size of a cooking pot.

The solar concentrator heating systems using parabolic mirrors or heliostats are now generating over 500°C for different applications purely from solar thermal energy thereby offsetting large quantities of carbon dioxide emission.

9) Trombe wall

It is a wall with high thermal mass used to store solar energy passively in a solar home. It is named after the French inventor, Felix Trombe, who popularized the design in 1964 although Edward Morse had patented it back in 1881.

A Trombe wall consists of a vertical wall, built of a material such as stone, concrete, or adobe, which is covered on the outside with glazing. Sunlight passing through the glazing generates heat which conducts through the wall. Warm air between the glazing and the Trombe wall surface can also be channeled by natural convection into the building interior or to the outside, depending on the building's heating or cooling needs.

During the day, sunlight shines through the glazing and hits the surface of the thermal mass, warming it by absorption. The air between the glazing and the thermal mass warms (via heat conduction) and rises, taking heat with it (convection). The warmer

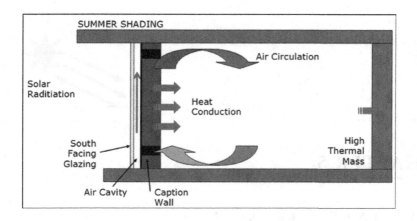

air moves through vents at the top of the wall and into the living area while cool air from the living area enters at vents near the bottom of the wall.

At night, a one-way flap on the bottom vent prevents backflow, which could act to cool the living area, and heat stored in the thermal mass radiates into the living area.

Advantages

- Comfortable Heat: Radiates in the infra red, which is more penetrating and pleasant than traditional convective forced air heating systems.
- Passive: No moving parts and essentially no maintenance.
- Simple Construction: Relatively easy to incorporate into building structure as an internal or external wall. Materials (masonry, concrete) are relatively inexpensive.
- Effective: Can reduce heating bills by large amounts.

Disadvantages

- Extended Overcast: Exterior walls become a heat loss source during extended overcast days. Not a problem for interior Trombe wall.

CHAPTER 5

CLIMATE CHANGE

> *To waste, to destroy our natural resources, to skin and exhaust the land instead of using it so as to increase its usefulness, will result in undermining in the days of our children the very prosperity which we ought by right to hand down to them amplified and developed.*
> *~Theodore Roosevelt, seventh annual message,*
> *3 December 1907*

Let us look at how experts define climate change.

NSIDC—Artic Climatology & Meteorology
Climate change—A study dealing with variations in climate on many different time scales from decades to millions of years, and the possible causes of such variations.

1) In the most general sense, the term "climate change" encompasses all forms of climatic inconstancy (that is, any differences between long-term statistics of the meteorological elements calculated for different periods but relating to the same area) regardless of their statistical nature or physical causes. Climate change may result from such factors as changes in solar activity, long-period changes in the Earth's orbital elements (eccentricity, obliquity of the ecliptic, precession of equinoxes), natural internal processes of the climate system, or anthropogenic forcing (for example,

increasing atmospheric concentrations of carbon dioxide and other greenhouse gases).
2) The term "climate change" is often used in a more restricted sense, to denote a significant change (such as a change having important economic, environmental and social effects) in the mean values of a meteorological element (in particular temperature or amount of precipitation) in the course of a certain period of time, where the means are taken over periods of the order of a decade or longer.

IPCC (Intergovernmental panel on Climate Change):

Climate change refers to a change in the state of the climate that can be identified (e.g. using statistical tests) by changes in the mean and/or the variability of its properties and that persists for an extended period, typically decades or longer. It refers to any change in climate over time, whether due to natural variability or as a result of human activity.

UNFCC (United Nations Framework on Climate Change):

Climate change refers to a change of climate that is attributed directly or indirectly to human activity that alters the composition of the global atmosphere and that is in addition to natural climate variability observed over comparable time periods.

Climate change is now affecting our daily life and through global warming threatens to escalate to more serious consequences for our future generations.

Radiative (Climate) forcing:

Radiative forcing is the influence of a factor such as greenhouse gas, which can cause climate change. It is the difference (measured in watts per sq meter) of solar energy received by earth and the amount of energy radiated back to space. This phenomenon is significant with its impact on the living and non living of this planet.

The word radiative arises because these factors change the balance between incoming solar radiation and outgoing infrared radiation within the Earth's atmosphere. This radiative balance controls the Earth's surface temperature. The term forcing is used to indicate that Earth's radiative balance is being pushed away from its normal state.

. To explain further, the earth is continually receiving energy from the sun. The earth loose energy to the space since space is much cooler than earth. Hence radiative forcing is the rate at which the surface of planet earth is heating or cooling.

IPCC in Fourth Assessment Report explains "Radiative forcing is a measure of how the energy balance of the Earth-atmosphere system is influenced when factors that affect climate are altered.

Radiative forcing is usually quantified as the 'rate of energy change per unit area of the globe as measured at the top of the atmosphere', and is expressed in units of 'Watts per square metre'. When radiative forcing from a factor or group of factors is evaluated as positive, the energy of the Earth-atmosphere system will ultimately increase, leading to a warming of the system. In contrast, for a negative radiative forcing, the energy will ultimately decrease, leading to a cooling of the system."

Global Warming Potential:

GWP is an estimate of how much a given green house gas will contribute to radiative forcing. Generally the GWP is specified over a period of 100 years. This means the effect of gas on radiative forcing over a century is the basis of any computation even though these gases will have different GWP in the short term. The GWP of carbon dioxide is 1, and a gas with GWP of 20 will the effect on radiative forcing by 20 times more than equal amount of carbon dioxide.

Equivalent carbon dioxide (CO_2e Kyoto):

CO_2e is an estimate of the concentration of CO_2 (in ppm) that would cause a given level of radiative forcing. The six major green house gases is accounted for and equalized to the effect of CO_2 on radiative forcing.

Total Equivalent carbon dioxide (CO_2e Total)

Total equivalent forcing is the forcings of all gases combined. This is also affected by the presence of sulphate aerosols, ozone and other cloud formations. Many among these others have a negative forcing impact and hence total radiative forcing is lesser than forcing of the Kyoto green house gases combined.

<u>**Green House gases:**</u>

1) Carbon dioxide (CO_2)
   Global Warming Potential—1

Major Sources:

- Burning of fossil fuel
- Deforestation
- Industrial production processes

2) Methane (CH_4)
Global Warming Potential—21
Major Sources:

- Livestock
- Waste in land fills
- Natural gas production

3) Nitrous Oxide (NO_2)
Global Warming Potential—310
Major Sources:

- Agriculture soil tilling
- Gasoline fuel
- Manufacturing of fertilizers

Green House Gas	Pre Industrial Level	Current level	Radiative Forcing (W/m^2)
Carbon dioxide	288 ppm	394 ppm	1.81
Methane	700 ppb	1745 ppb	0.51
Nitrous Oxide	270 ppb	323 ppb	0.18
CFC—12	0	533 ppt	0.17

4) Hydro flouro carbons (HFCs)
Global Warming Potential—1200 to 12000
Major Sources:

- Cooling fluid of air conditioners
- Thermal insulation for utilities
- Ingredients of cosmetics

Other major pollutants:

Black Carbon

Black carbon (BC) exists as particles in the atmosphere and is a major component of soot. BC is not a greenhouse gas. Instead it warms the atmosphere by intercepting sunlight and absorbing it. BC and other particles are emitted from many common sources, such as cars and trucks, residential stoves, forest fires and some industrial facilities. BC particles have a strong warming effect in the atmosphere, darken snow when it is deposited, and influence cloud formation. These anthropogenic particles are also known to have a negative impact on human health.

Black carbon results from the incomplete combustion of fossil fuels, wood and other biomass. Complete combustion would turn all carbon in the fuel into carbon dioxide (CO_2). In practice, combustion is never complete and CO_2, carbon monoxide (CO), volatile organic compounds (VOCs), organic carbon (OC) particles and BC particles are all formed. There is a close relationship between emissions of BC (a warming agent) and OC (a cooling agent). They are always co-emitted, but in different proportions for different sources.

The black in BC refers to the fact that these particles absorb visible light. This absorption leads to a disturbance of the planetary radiation balance and eventually to warming. The contribution to warming of 1 gram of BC seen over a period of 100 years has been estimated to be anything from 100 to 2000 times higher than that of 1 gram of CO_2. An important aspect of BC particles is that their lifetime in the atmosphere is short, days to weeks, and so emission reductions have an immediate benefit for climate and health.

Tropospheric Ozone:

Ozone (O_3) is a reactive gas that exists in two layers of the atmosphere: the stratosphere (the upper layer) and the troposphere (ground level to ~10-15 km). In the stratosphere, O_3 is considered to be beneficial as it protects life on Earth from the sun's harmful ultraviolet (UV) radiation. In contrast, at ground level, it is an air pollutant harmful to human health and ecosystems, and it is a major component of urban smog. In the troposphere, O_3 is also a significant greenhouse gas.

The three fold increase of the O3 concentration in the northern hemisphere during the past 100 years has made it the third most important contributor to the human enhancement of the global greenhouse effect, after CO2 and CH4.

In the troposphere, O3 is formed by the action of sunlight on O3 precursors that have natural and anthropogenic sources. These precursors are CH4, nitrogen oxides (NOX), VOCs and CO. Reductions in both CH4 and CO emissions have the potential to substantially reduce O3 concentrations and reduce global warming while for NOX and VOC the impact on climate change is minimal.

Symptoms and Effects of climate change:

The climate change is causing very serious damage to our mother earth and it is manifested through various phenomena including

1) Global Warming
2) Green House Effect
3) Urban smog
4) Urban heat island effect
5) Acid rain
6) Ozone Depletion
7) Impact on continents
8) Impact on ecosystem
9) Impact on biodiversity
10) Impact on agriculture

Global warming:

Global warming is the increase in the average temperature of Earth's near-surface air and oceans since the mid-20th century and its projected continuation. According to the Intergovernmental Panel on Climate Change (IPCC), global surface temperature increased 0.8°C during the 20th century. Most of the observed temperature increase since the middle of the 20th century has been caused by increasing concentrations of greenhouse gases, which result from human activity such as the burning of fossil fuel and deforestation. Climate model projections summarized in the latest IPCC report indicate that the global surface temperature is likely to rise a further 1.1°C and above during the 21st century.

Effects of global warming:

- Cause sea levels to rise
- Change the amount and pattern of precipitation
- Expansion of subtropical deserts
- Continuing retreat of glaciers, permafrost and sea ice
- Changes in the frequency and intensity of extreme weather events
- Species extinctions
- Changes in agricultural yields
- Ocean acidification

Average temperature movement:

Evidence for warming of the climate system includes observed increases in global average air and ocean temperatures, widespread melting of snow and ice, and rising global average sea level. The most common measure of global warming is the trend in globally averaged temperature near the earth's surface

Year	1850	1860	1870	1880	1890	1900	1910	1920
Avg Temp(°C)	13.7	13.6	13.8	13.7	13.8	13.7	13.5	13.6
Year	1930	1940	1950	1960	1970	1980	1990	2000
Avg Temp(°C)	13.7	14	13.8	13.9	14	14.1	14.3	14.5

Source: IPCC Third assessment report

Earth's surface temperature rose by about 0.8 °C over the period 1900-2000. The rate of warming over the last half of that period was almost double that for the period as a whole. This is frightening as this temperature is believed to be relatively stable over one or two thousand years before 1850, with only variations (Medieval Warm Period, Little Ice Age etc) regionally. The industrial revolution in the 19th & 20th century has had its effect on the climate. More so the second revolution starting about 1850 and getting pronounced in the early 1900—with expansion in coal based power, automobiles with fossil fuel, conflicts resulting in war between nations and group of nations, deforestation in the name of development etc all have contributed to the temperature increase. Years 1998 and 2005 are the warmest years.

Temperature changes vary around the globe. Since 1979, land temperatures have increased about twice as fast as ocean temperatures (0.25 °C per decade against 0.13 °C per decade). Ocean temperatures increase more slowly than land temperatures because of the larger effective heat capacity of the oceans and because the ocean loses more heat by evaporation.

Due to high thermal inertia of oceans and its combination with slow response of other indirect effects makes it more difficult to fully control global warming. Studies have shown that even if green house gases were stabilised at year 2000 levels, a further warming of 0.5°C would still occur.

Increases in sea level are consistent with warming. Studies have confirmed global average sea level rose at an average rate of 1.8 mm per year over 1961 to 2003 and at an average rate of about 3.1 mm per year from 1993 to 2003.

The sea level increases due to

(1) Thermal expansion of oceans
(2) Decrease in glaciers and icecaps
(3) Losses from the polar ice sheets.

Greenhouse Effect:
. Of the visible sunlight that enters the atmosphere, about 30% is reflected back into space by clouds, snow and ice-covered land, sea surfaces, and atmospheric dust. The rest is absorbed by the liquids, solids, and gases in the atmosphere and on the earth surface. The excess energy absorbed is reemitted as infrared radiation. This radiation is partly trapped by naturally occurring water vapour and other particles in the atmosphere. This in turn increases the atmospheric temperature. This warming is termed as green house effect.

It is established that the naturally occurring greenhouse gases have a mean warming effect of about 33 °C which is required for the existence of living organism. A delicate balance is maintained here by the nature for its sustenance. This balance is now being tilted towards a danger zone with the increase in anthropogenic greenhouse gases such as carbon dioxide (CO_2), methane (CH_4), ozone (O_3) etc resulting in excessive green house effect.

Human activity since the Industrial Revolution has increased the amount of anthropogenic greenhouse gases in the atmosphere.

The concentrations of CO_2 and methane have increased considerably since pre industrial times. These levels are known to be much higher than at any time during the last 500,000 years. Fossil fuel burning has produced about three-quarters of the increase in CO_2 from human activity over the past 20 years. Most of the rest is due to land-use change, particularly deforestation.

Urban smog:

Urban smog, which is known as ozone pollution, is produced by a complex series of chemical reactions involving automotive and industrial emissions of volatile organic compounds (VOCs, mainly hydrocarbons), nitrogen oxides (NOx) from the same sources, and sunlight. As temperatures increase during the day, solar energy enhances those chemical reactions and increases the amount of ozone produced. Correspondingly, as temperatures decrease, the chemical reactions are slowed and smog is seldom formed. Ozone formation here is thus a daytime phenomenon.

Large urban centers have compounded chemical reactions from the interaction of sunlight with hydrocarbons and nitrogen oxides, mostly from car exhaust.

Emissions from industrial facilities and electric utilities, motor vehicle exhaust, gasoline vapors, and chemical solvents are all major sources of NOx and VOC.

Breathing ozone, a primary component of smog, can trigger a variety of health problems including chest pain, coughing, throat irritation, and congestion. It can worsen bronchitis, emphysema, and asthma. Ground-level ozone also can reduce lung function and inflame the linings of the lungs. Repeated exposure may permanently scar lung tissue.

Ground-level ozone also damages vegetation and ecosystems. In the United States alone, ozone is responsible for an estimated $500 million in reduced crop production each year.

Urban heat island effect:

Urban heat island is the rise in temperature of any man made area resulting in a well defined distinct warm island among the cooler nearby lower temperature areas. Around half of the world's human population lives in urban areas. In the near future it is expected that the global rate of urbanization will increase by 70% of the present world urban population by 2030, as urban agglomerations emerge and population migration from rural to urban/suburban areas continues.

Urbanization negatively impacts the environment mainly by the production of pollution, the modification of the physical and chemical properties of the atmosphere, and the covering of the soil surface. The cumulative effect of all these impacts is the UHI. UHI is formed in cities, since their surfaces are prone to release large quantities of heat. Nonetheless, the UHI negatively impacts not only residents of urban-related environs, but also humans and their associated ecosystems located far away from cities. In fact, UHIs have been indirectly related to climate change due to their contribution to the greenhouse effect, and therefore, to global warming.

Acid rain:

"Acid rain" is a broad term referring to a mixture of wet and dry deposition from the atmosphere containing higher than normal amounts of nitric and sulfuric acids. The acid rain formation result from both natural sources, such as volcanoes and decaying vegetation, and man-made sources, primarily emissions of sulfur dioxide (SO_2) and nitrogen oxides (NO_x) resulting from fossil fuel combustion. SO_2 and NO_x largely come from electric power generation that relies on burning fossil fuels, like coal. Acid rain occurs when these gases react in the atmosphere with water, oxygen, and other chemicals to form various acidic compounds. The result is a mild solution of sulfuric acid and nitric acid. The acid rain need not always occur near the generation points of these gases. When sulfur dioxide and nitrogen oxides are released from power plants and other sources, prevailing winds blow these compounds across state and national borders, sometimes over hundreds of miles.

Wet Deposition

Wet deposition refers to acidic rain, fog, and snow. If the acid chemicals in the air are blown into areas where the weather is wet, the acids can fall to the ground in the form of rain, snow, fog, or mist. As this acidic water flows over and through the ground, it affects a variety of plants and animals. The strength of the effects depends on several factors, including how acidic the water is; the chemistry and buffering capacity of the soils involved; and the types of fish, trees, and other living things that rely on the water.

Dry Deposition

In areas where the weather is dry, the acid chemicals may become incorporated into dust or smoke and fall to the ground through dry deposition, sticking to the ground, buildings, homes, cars, and trees. Dry deposited gases and particles can be washed from these surfaces by rainstorms, leading to increased runoff. This runoff water makes the resulting mixture more acidic. About half of the acidity in the atmosphere falls back to earth through dry deposition.

Acid rain causes acidification of lakes and streams and contributes to the damage of trees at high elevations (for example, red spruce trees above 2,000 feet) and many sensitive forest soils. In addition, acid rain accelerates the decay of building materials and paints, including irreplaceable buildings, statues, and sculptures that are part of our nation's cultural heritage. Prior to falling to the earth, sulfur dioxide (SO_2) and nitrogen oxide (NO_x) gases and their particulate matter derivatives—sulfates and nitrates—contribute to visibility degradation and harm public health.

Ozone Depletion:

Stratospheric Ozone protects all living organism from the harmful Ultra violet rays by absorbing large part of Ultra Violet—B coming from the sun. Excessive depletion of this Ozone will be devastating.

Ozone depletion describes two distinct but related phenomena observed since the late 1970s: a steady decline of about 4% per decade in the total volume of ozone in Earth's stratosphere (the ozone layer), and a much larger yearly springtime decrease in stratospheric ozone over Earth's polar regions. The latter phenomenon is referred to as the ozone hole.

The details of polar ozone hole formation differ from that of mid-latitude thinning, but the most important process in both is catalytic destruction of ozone by atomic halogens. The main source of these halogen atoms in the stratosphere is photo dissociation of man-made Halocarbon refrigerants (CFCs, freons, halons.) These compounds are transported into the stratosphere after being emitted at the surface. Both types of ozone depletion were observed to increase as emissions of Halocarbons increased. CFCs and other contributory substances are referred to as **ozone-depleting substances** (**ODS**). The Montreal protocol controls and its implementation have substantially reduced the emission of ODS chemicals.

Antarctic ozone hole

The discovery of the Antarctic "ozone hole" by British Antarctic Survey scientists Farman, Gardiner and Shanklin (announced in a paper in *Nature* in May 1985) came as a shock to the scientific community, because the observed decline in polar ozone was far larger than anyone had anticipated. Satellite measurements showing massive depletion of ozone around the South Pole were becoming available and a detailed analysis showed the ozone hole was existing as far back as 1976. This is expected to have been contained to a large extend by the good work done.

The 2007 UNEP sponsored report showed that the hole in the ozone layer was recovering and the smallest it had been for about a decade. The 2010 report found that over the past decade, global ozone and ozone in the Arctic and Antarctic regions is no longer decreasing but is not yet increasing. The ozone layer outside the Polar Regions is projected to recover to its pre-1980 levels some time before the middle of this century and the springtime ozone hole over the Antarctic is expected to recover much later.

Arctic ozone hole

The ozone depletion around the Artic region happens during January to March. The significant loss ozone layer was seen above the Artic in early 2011. March recorded only half of the ozone layer present around the pole. This dilution of ozone layer is caused by the climate change impact. Studies are pointing to the fact that excessive trapping of infrared radiation by green house gases present in the lower atmosphere is cooling the Stratosphere thereby aiding higher ozone disintegration.

Further dilution of the ozone layer will result in a creation of ozone hole over Artic region. Unlike the Antarctica Ozone Hole, any ozone hole formation over and around the Artic Pole will expose millions of people, animals vegetation etc and hence more serious. However this has recovered to a large extend, but a prospect of a dangerous ozone hole looms large.

Tibet ozone hole

As winters that are colder are more affected, at times there is an ozone hole over Tibet. In 2006, an ozone hole was detected over Tibet. Also again in 2011 an ozone hole appeared over mountains of Tibet, Xinjiang, Qinghai and the Hindu Kush, along with an unprecedented hole over the Arctic, though the Tibet one is far less intense than the ones over the arctic or antarctic.

Impact on continents:

Africa

About 8 to 10 years from now, between 100 million of people are projected to be exposed to increased water stress due to climate change. Yields from rain-fed agriculture could be reduced significantly. Agricultural production, including access to food, in many African countries is projected to be severely compromised. This would further adversely affect food security and exacerbate malnutrition. Towards the end of the 21st century, projected sea level rise will affect low-lying coastal areas with large populations. The cost of adaptation could amount to at least 10% of GDP which is not affordable to many countries in this region.

Asia

By the middle of the century, freshwater availability in Central, South, East and South-East Asia, particularly in large river basins, is projected to decrease. Coastal areas, especially heavily populated mega delta regions in South, East and South-East Asia, will be at greatest risk due to increased flooding from the sea and, in some mega deltas, flooding from the rivers. Climate change is projected to compound the pressures on natural resources and the environment associated with rapid urbanization, industrialization and economic development. Diarrhoeal disease primarily associated with floods and droughts are expected to rise in East, South and South-East Asia due to projected changes in the hydrological cycle.

Australia and New Zealand

In the next 10 to 20 years significant loss of biodiversity is projected to occur in some ecologically rich sites, including the Great Barrier Reef and Queensland Wet Tropics. Water security problems are projected to intensify in southern and eastern Australia and, in New Zealand, in Northland and some eastern regions. The production from agriculture and forestry is projected to decline over much of southern and eastern Australia, and over parts of eastern New Zealand, due to increased drought and fire. By middle of the century, ongoing coastal development and population growth in some areas of Australia and New Zealand are projected to exacerbate risks from sea level rise and increases in the severity and frequency of storms and coastal flooding.

Europe

Climate change is expected to magnify regional differences in Europe's natural resources and assets. Negative impacts will include increased risk of inland flash floods and more frequent coastal flooding and increased erosion (due to storminess and sea level rise). Mountainous areas will face glacier retreat, reduced snow cover and winter tourism, and extensive species losses.

In southern Europe, climate change is projected to worsen conditions (high temperatures and drought) in a region already vulnerable to climate variability, and to reduce water availability, hydropower potential, summer tourism and, in general, crop productivity. Climate change is also projected to increase the health risks due to heat waves and the frequency of wildfires.

Latin America

By mid-century, increases in temperature and associated decreases in soil water are projected to lead to gradual replacement of tropical forest by savanna in eastern Amazonia. Semi arid vegetation will tend to be replaced by arid-land vegetation. There is a risk of significant biodiversity loss through species extinction in many areas of tropical Latin America. Productivity of some important crops is projected to decrease and livestock productivity to decline, with adverse consequences for food security. The number of people at risk of hunger is projected to increase. Changes in precipitation patterns and the disappearance of glaciers are projected to significantly affect water availability for human consumption, agriculture and energy generation.

Impact on ecosystem:

Systems

The adaptability of many ecosystems will be under strain by an unprecedented combination of climate change, associated disturbances (e.g. flooding, drought, wildfire, insects, ocean acidification) and other global change drivers (e.g. land use change, pollution, fragmentation of natural systems, overexploitation of resources).Over the course of this century, net carbon uptake by terrestrial ecosystems is likely to peak before mid-century and then weaken or even reverse, thus amplifying climate change. Approximately 20 to 30% of plant and animal species assessed so far are likely to be at increased risk of extinction if increases in global average temperature exceed 1.5 to 2°C. Negative consequences for biodiversity and ecosystem goods and services, e.g. water and food supply are a direct effect of increase in global average temperature and CO_2 concentration.

Food

Crop productivity is projected to increase slightly at mid—to high latitudes for local mean temperature increases of up to 1 to 3°C depending on the crop, and then decrease beyond that in some regions. At lower latitudes, especially in seasonally dry and tropical regions, crop productivity is projected to decrease for even small local temperature increases (1 to 2°C), which would increase the risk of hunger. The unforeseen flood and drought will destroy crop yields and severely affect the food chain. The rising sea salt water is a major threat to food grain cultivation in coastal areas.

Coasts

Coasts are projected to be exposed to increasing risks, including coastal erosion, due to climate change and sea level rise. The effect will be exacerbated by increasing human-induced pressures on coastal areas. Sand mining and non biodegradable waste dumping near coastal areas are causing serious damage to the adjoining land and sea system. It is already a delicate balance now. This will be tilted and catastrophic floods affecting millions may result with rise in sea level.

Industry, settlements and society

The most vulnerable industries, settlements and societies are generally those in coastal and river flood plains, those whose economies are closely linked with climate-sensitive resources and those in areas prone to extreme weather events, especially where rapid urbanization is occurring. Poor communities can be especially vulnerable, in particular those concentrated in high-risk areas.

Health

The health status of millions of people is projected to be affected through increase in malnutrition; diseases and injury due to extreme weather events; increased burden of diarrhoeal diseases; increased frequency of cardio-respiratory diseases due to higher concentrations of ground-level ozone in urban areas related to climate change; and the altered spatial distribution of some infectious diseases.

Climate change is projected to bring some benefits in low temperate areas, such as lower intensity of cold exposure, and some mixed effects such as changes in range and transmission potential of malaria in Africa. However overall the benefits will be outweighed by the negative health effects of rising temperatures, especially in developing countries.

Impact on Bio diversity:

The variety of life on Earth, its biological diversity is commonly referred to as biodiversity. The number of species of plants, animals, and micro organisms, the enormous diversity of genes in these species, the different ecosystems on the planet, such as deserts, rainforests and coral reefs are all part of a biologically diverse earth.

Climate change is already having an impact on biodiversity, and is projected to become a progressively more significant threat in the coming decades. The related pressure of ocean acidification, resulting from higher concentrations of carbon dioxide in the atmosphere, is also already being observed. In addition to warming temperatures, more frequent extreme weather events and changing patterns of rainfall and drought can be expected to have significant impacts on biodiversity. Further effects include

- Prospect of ice free summers in the Artic
- When Ice is replaced with water heat absorption by ocean accelerate
- Ocean acidification—more CO_2 makes the ocean acidic by formation of carbonic acid
- Ocean stratification occurs and affects marine organism Phytoplankton and on lizards
- Coral reefs are threatened

Impact on Agriculture:

The impacts of climate change on agriculture and human well-being include:

1) The biological effects on crop yields;
2) The resulting impacts on outcomes including prices, production, and consumption
3) The impacts on per capita calorie consumption and child malnutrition.

The biophysical effects of climate change on agriculture induce changes in production and prices, which play out through the economic system as farmers and other market participants adjust autonomously, altering crop mix, input use, production, food demand, food consumption, and trade. Rising temperatures and changes in rainfall patterns have direct effects on crop yields, as well as indirect effects through changes in irrigation water availability.

Climate change is a reality now and this need to be addressed with utmost urgency. Two pronged strategies are adopted world wide to reduce the adverse impact of climate change. 1) Mitigation 2) Adaptation

Mitigation

Reducing the amount of future climate change is called mitigation of climate change. The IPCC defines mitigation as activities that reduce greenhouse gas (GHG) emissions, or enhance the capacity of carbon sinks to absorb GHGs from the atmosphere. Many countries, both developing and developed, are aiming to use cleaner, less polluting, technologies. Use of these technologies aids mitigation and could result in substantial reductions in CO_2 emissions. Many countries are adopting policies that include targets for emissions reductions, increased use of renewable energy, and increased energy efficiency.

Techniques/ Activities to support mitigation include

1. Use of renewable power from solar, wind, biogas etc
2. Protect natural carbon sinks like forests and oceans
3. Create and use more energy efficient products
4. Rapidly increase the usage of bicycles

5. Improve insulation and efficiency in buildings
6. Increase use and development of low carbon technologies
7. Implement stricter norms in fossil fuel emissions
8. Increase the use of all electric/ hybrid vehicles
9. Use of biofuels for transport and industry
10. Increase crop yields

Adaptation

Climate change adaptation (CCA) is a process in which the vulnerable society is advised, guided and supported on the impending adverse effects on health and well being due to climate change through economic and social initiatives. Adaptation to climate change may be planned, e.g., by local or national government, or spontaneous, i.e., done privately without government intervention. Even if the emissions are controlled on a war footing, global warming will continue for some more time and will affect the vulnerable communities. The ability to adapt is closely linked to social and economic development. Since these communities mostly lie in the bottom of the economic pyramid, the climate change adaptation need to be aggressively pursued to support them. Adaptation gains more relevance with the fact that all climate change cannot be mitigated. The barriers, limits, and costs of future adaptation need to be understood more.

Techniques/ Activities to support adaptation include:

1. Change architecture of buildings and establish new codes
2. Installing early warning systems
3. Urban planning and preventive vaccinations
4. Constructing protective wall in sea cost
5. Migration to a safer location
6. Decongestion of high risk areas
7. Creation of a good resource management system
8. Soil conservation and drought tolerant crops cultivation
9. Use drip irrigation and grass waterways
10. Change farming practices and rotate crops

Of the above it is through mitigation that we can effectively delay and stall the effect of climate change on the inhabitants. However mitigation and adaptation are not alternatives. Now they both need to be pursued actively in parallel to overcome the negative impact of climate change.

International cooperation meetings on climate change:

Montreal Protocol:

The Montreal Protocol on Substances that deplete the Ozone Layer (a protocol to the Vienna Convention for the Protection of the Ozone Layer) is an international treaty designed to protect the ozone layer by phasing out the production of numerous substances believed to be responsible for ozone depletion. The treaty was opened for signature on September 16, 1987, and entered into force on January 1, 1989, followed by a first meeting in Helsinki, May 1989. Since then, it has undergone seven revisions. If the international agreement is adhered to, the ozone layer is expected to recover by 2050. This protocol which was adopted and implemented throughout the world is an example of international cooperation to combat man made disasters. This protocol restricted the production and use of halogenated hydrocarbons, specifically those containing chlorine or bromine—the chemical that have adverse effect on the Ozone layer.

The protocol laid down the terms to phase out

- Chlorofluoro carbons(CFCs)in developed countries by 1996
- Chlorofluoro carbons(CFCs)in developing countries by 2010
- Hydrochlorofluoro carbons(HCFCs) by 2030

Due to the Montreal Protocol, the atmospheric concentrations of the most important chlorofluorocarbons and related chlorinated hydrocarbons have either leveled off or decreased. Halon concentrations have continued to increase, as the halons presently stored in fire extinguishers are released, but their rate of increase has slowed and their abundances are expected to begin to decline by about 2020. Also, the concentration of the HCFCs increased drastically at least partly because for many uses of CFCs (e.g. used as solvents or refrigerating agents) were substituted with HCFCs. The overall level of compliance has been high.

Website: http://www.unep.org/ozone/montreal/

Kyoto Protocol:

The Kyoto Protocol is an international agreement linked to the United Nations Framework Convention on Climate Change. The major feature of the Kyoto Protocol is that it sets binding targets for 37 industrialized countries and the European community for reducing greenhouse gas (GHG) emissions. These amount to an average of five per cent against 1990 levels over the five-year period 2008-2012.

The major distinction between the Protocol and the Convention is that while the Convention encouraged industrialized countries to stabilize GHG emissions, the Protocol commits them to do so.

Recognizing that developed countries are principally responsible for the current high levels of GHG emissions in the atmosphere as a result of more than 150 years of industrial activity, the Protocol places a heavier burden on developed nations under the principle of "common but differentiated responsibilities."

The Kyoto Protocol was adopted in Kyoto, Japan, on 11 December 1997 and entered into force on 16 February 2005. The detailed rules for the implementation of the Protocol were adopted at COP 7 in Marrakesh in 2001, and are called the "Marrakesh Accords."

The Kyoto mechanisms

Under the Treaty, countries must meet their targets primarily through national measures. However, the Kyoto Protocol offers them an additional means of meeting their targets by way of three market-based mechanisms.

The Kyoto mechanisms are:

- Emissions trading—known as "the carbon market"
- Clean development mechanism (CDM)
- Joint implementation (JI).

The mechanisms help stimulate green investment and help Parties meet their emission targets in a cost-effective way.
Website: http://unfccc.int/kyoto_protocol

Copenhagen Accord:

The Copenhagen Accord is a document that delegates at the 15th session of the Conference of Parties (COP 15) to the United Nations

Framework Convention on Climate Change agreed to "take note of" at the final plenary on 18 December 2009.

The Accord

- Endorses the continuation of the Kyoto Protocol.
- Underlines that climate change is one of the greatest challenges of our time and emphasizes a "strong political will to urgently combat climate change in accordance with the principle of common but differentiated responsibilities and respective capabilities"
- To prevent dangerous anthropogenic interference with the climate system, recognizes "the scientific view that the increase in global temperature should be below 2 degrees Celsius", in a context of sustainable development, to combat climate change.
- Recognizes "the critical impacts of climate change and the potential impacts of response measures on countries particularly vulnerable to its adverse effects" and stresses "the need to establish a comprehensive adaptation programme including international support"
- Recognizes that "deep cuts in global emissions are required according to science" (IPCC AR4) and agrees cooperation in peaking (stopping from rising) global and national greenhouse gas emissions "as soon as possible" and that "a low-emission development strategy is indispensable to sustainable development"
- States that "enhanced action and international cooperation on adaptation is urgently required
- About mitigation agrees that developed countries would "commit to economy-wide emissions targets for 2020" to be submitted by 31 January 2010
- Agrees that developing nations would "implement mitigation actions" (Nationally Appropriate Mitigation Actions) to slow growth in their carbon emissions, submitting these by 31 January 2010.
- Agrees that developing countries would report those actions once every two years
- Recognizes "the crucial role of reducing emission from deforestation and forest degradation and the need to enhance removals of greenhouse gas emission by forests",

http://unfccc.int/meetings/cop_15/copenhagen_accord/items/5262.php

RIO + 20:

Background

The United Nations Conference on Sustainable Development (UNCSD) is being organized in pursuance of General Assembly Resolution 64/236. The Conference took place place in Brazil on 4-6 June 2012 to mark the 20th anniversary of the 1992 United Nations Conference on Environment and Development (UNCED), in Rio de Janeiro, and the 10th anniversary of the 2002 World Summit on Sustainable Development (WSSD) in Johannesburg. It is envisaged as a Conference at the highest possible level, including Heads of State and Government or other representatives.

Objective of the Conference

The objective of the Conference is to secure renewed political commitment for sustainable development, assess the progress to date and the remaining gaps in the implementation of the outcomes of the major summits on sustainable development, and address new and emerging challenges

Themes of the Conference

The Conference will focus on two themes: (a) a green economy in the context of sustainable development and poverty eradication; and (b) the institutional framework for sustainable development.

Outcome document: The outcome document for this conference is titled "The Future We Want". This 50 odd page document have 280 odd points. It sought to address the global sustainable development and poverty alleviation under the following headings

1. Our common vision
2. Renewing political commitment
3. Green economy in the context of sustainable development and poverty eradication
4. Institutional fame work for sustainable development
5. Framework for action and followup
6. Means of implementation

Website: http://www.uncsd2012.org
Source: Respective Websites

A few organisations working on climate change:

1) The World Bank
   Website: http://beta.worldbank.org/climatechange/
2) Intergovernmental Panel on Climate Change (IPCC)
   Website: www.ipcc.ch/index.htm
3) United Nations Framework on Climate Change (UNFCC):
   Website: http://unfccc.int
4) The International Union for Conservation of Nature (IUCN)
   Website: www.iucn.org
5) United Nations Environmental Programme (UNEP)
   Website: www.unep.org
6) International Energy Agency
   Website: www.iea.org
7) Both ENDS
   Website: www.bothends.org/
8) Green Peace International:
   Website: www.greenpeace.org/international/en/
9) World Wide Fund
   Website: www.panda.org
10) German Watch
    Website: www.germanwatch.org
11) Friends of Earth International:
    Website: www.foei.org/
12) World Future Council
    Website: www.worldfuturecouncil.org
13) 350.org
    Website: www.350.org
14) NSIDC (Artic climatology & Meteorology)
    Website: http://nsidc.org/arcticmet/
15) Conservation International
    Website: www.conservation.org
16) Centre for Climate and Energy Solutions
    Website: www.c2es.org
17) Climate Institute
    Website: www.climateinstitute.org.au
18) European Commission
    Website: http://ec.europa.eu/dgs/clima/mission/index_en.htm

19) Climate Action Network (CAN)
   Website: www.climatenetwork.org
20) Green Cross International:
   Website: http://www.gci.ch/
21) The 3C initiative;
   Website: www.combactclimatechange.org
22) WBCSD:
   Website: www.wbcsd.org
23) Copenhagen climate council
   Website: http://www.copenhagenclimatecouncil.com
24) The climate group
   Website: www.theclimategroup.org
25) ISEO
   Website: www.uniseo.org

The entire list of the organizations working on climate change will be very exhaustive. Hence this list may not contain many. Any omission is unintentional.

CHAPTER 6

SOLAR ENERGY—GLOBAL SCENARIO

"The great work now . . . . is to carry out the transition from a period of human devastation of the earth to a period when humans would be present to the planet in a mutually beneficial manner"

Thomas Berry

CIS tower Manchester _UK—Building integrated PV

Credit: Stephen Richards

- Renewable energy accounted for approximately half of the estimated 208GW of new electric capacity added globally in 2011
- Global renewable power capacity excluding hydro reached 390 GW in 2011
- Global Solar PV installed capacity reached 70 GW in 2011
- Global concentrating solar thermal power capacity reached 1.8GW by 2011
- Global Solar water heating capacity reached 232GWth in 2011 by adding 50 GWth during the year
- Germany, Italy & Japan ranked 1,2 &3 in Solar PV cumulative installed capacity in 2011
- China, Turkey & Germany ranked 1,2 &3 in Solar hot water generation capacity in 2010
- Worldwide over 1.5 million people are employed directly or indirectly in Solar Industry in 2011
- Global average PV module prices dropped 23% from $4.75/W in 1998 to $3.65/W in 2008 and to $0.9/W in 2011.
- In 2011, global spending on R&D for renewables was USD 8.3 billion in which 4 billion went to Solar
- The total investment in renewables during 2011 was a staggering USD 257 billion with investment in solar accounting for 147 billion.
- The global PV industry has seen impressive growth rates in cell/module production during the past decade, with a 10-year compound annual growth rate (CAGR) of 46% and a 5-year CAGR of 56% through 2008
- Thin-film PV technologies have grown faster than crystalline silicon over the past 5 years, with a 10-year CAGR of 47% and a 5-year CAGR of 87% for thin-film shipments through 2008. This has slowed down a bit in the last couple of years.
- Year 2010 saw the investment in renewables in developing countries surpass that of developed countries
- About 120 countries around the world have announced support for renewable energy harnessing and laid out policy target. Of this 50 are developing countries

Reference: NREL Market report_Jan2010
REN21_GSR_2012

Global Energy & Emission trend:

Commensurate to the growth in GDP and consumption worldwide, the energy consumption also grew. From around 50PentaWh in 1970 it reached about 100PentaWh in 2010. This show doubling the energy consumption took about 40 years. But now with high population countries experiencing good growth, the energy consumption is expected to double in next 20 to 25 years.

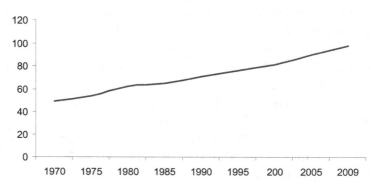

Similar to the final energy consumption, the worldwide electricity consumption increased drastically. From about 4.5PentaWh, during the last four decades the electricity consumption increased to over 17PentaWh. This is a four fold increase. This increase is pretty steep and shows the electricity consumption is growing at a much faster rate than the total energy consumption. The electricity consumption is expected to double the present level in the next 10 to 12 years.

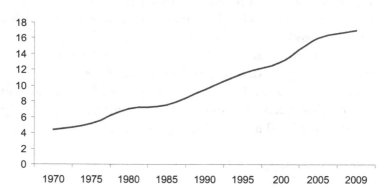

Since the electricity and energy is primarily depending on fossil fuels, the related emission cannot lag behind. The CO2 emission saw a high increase

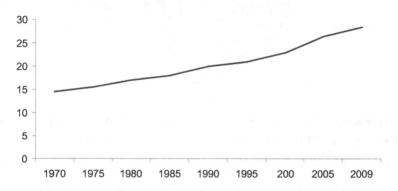

CO₂ Emission (Billion Tonnes)

During the period 1970 to 2010 from 14.5 billion tones it rose to about 30 billion tones. This unabated emission rate should be curtailed immediately to avoid catastrophic environmental disasters. The renewable energy sources which accounts for 20% of electricity production today need to be increased on a war footing with more emphasis on Wind and Solar

2010 Global Primary Energy Supply by Fuel

Fuel	MTOE	Penta Wh
Oil	4133	48
Coal	3434	40
Gas	2734	32
Bio energy	1272	15
Nuclear	725	8
Hydro	250	3
Solar & Wind Major	127	1

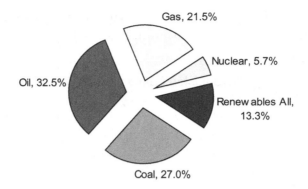

Renewables now contribute to 13% of global primary energy supply and expected to move up cautiously—Renewables all include Hydro, Biomass, Solar, Wind etc

Growth of share of renewables (With Solar and Wind major)

Year	1980	2000	2007	2010	2015	2030
World Primary Energy Demand (MTOE)	7228	10018	12013	12717	13488	16790
Renewables (MTOE)	12	55	74	127	160	370
% share	0.17%	0.55%	0.62%	1.06%	1.19%	2.20%

Reference:IEA WEO 2010

Renewables power capacity_ 2011

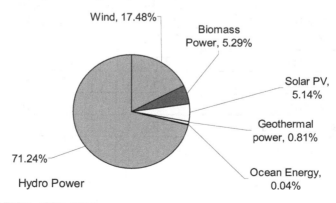

Reference: REN21_GSR_2012

Hydro Power is the leader now with 1010 GW and Wind energy comes second with a capacity of 198 GW followed at a distance by Biomass Power. Solar PV grid is at forth position but is growing fast. China topped the list of renewable energy (Excluding hydro power) capacity, followed closely by United States. The other countries among the top five are Germany, Spain and Italy at the 5th Position.

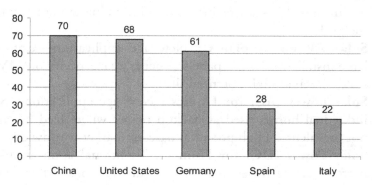

Reference: REN21_GSR_2012

Global Annual Investment in new renewable energy capacity

With hardly 20 billion in 2004 it increased to a staggering 257 billion dollars in 2012. The top countries for total investment in 2011 were China, Germany, the United States, Italy, and Brazil. For the first time in 2010, financial new investment in renewable energy in

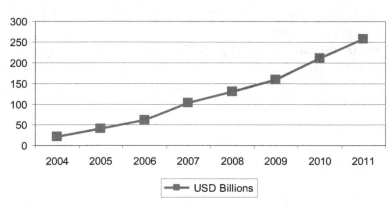

Reference: REN21_GSR_2012

developing countries surpassed that in developed economies rising from $17 billion to more than $72 billion, while in OECD countries it increased less than $4 billion to $70.5 billion

The Solar Scenario:
The world wide major application areas can be categorized as

- **Solar PV Grid connected :** New age application—Primarily seen in Europe—PV systems installed in individual homes and feeding the grid
- **Solar PV power plants :** New age application—Commercial large installation of PV panels to generate electricity on MW/GW scale—Seen at high potential sites
- **Solar power (AC) for Homes :** New age application—Seen in individual homes used as backup power where feed in tariff are not prevalent—High potential nascent market worldwide
- **Solar Power (DC) for Homes :** Traditional application—Seen more in unelectrified regions around the world—Business built on subsidies—But highly efficient systems
- **Solar thermal—Household, Industrial :** Traditional application—Used by individual homes / establishments where there is genuine need for hot water—Ideal replacement for electric water heater—seen around the world in urban and semi urban households
- **Solar Thermal—Power plants :** New age application—More known as CSP—Heat water at temp higher than 100 deg centigrade to produce steam using solar, thereby generating electricity—Very promising high focus area with lot for research being undertaken
- **Solar thermal—Heating & Cooling:** New age application—Apart from water, air is heated with solar heat. Additionally Air cooling is done by incorporating absorption refrigeration using solar heat.

Solar has grown impressively in the last few years

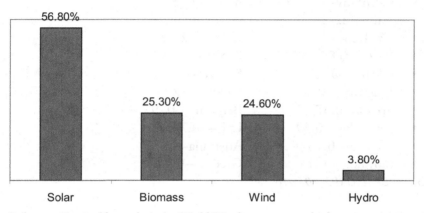

Reference: Renewable revolution—World Watch

Evolution of world annual PV market (MW)

2000—276 MW	2006—1596 MW
2001—334 MW	2007—2594 MW
2002—477 MW	2008—6090 MW
2003—583 MW	2009—7257 MW
2004—1122 MW	2010—16629 MW
2005—1422 MW	2011—29665 MW

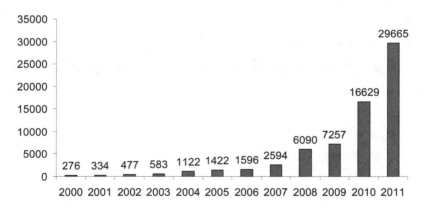

Reference: EPIA: GMO_2016

Annual PV installation in GW—Top Ten

2010	2011
• Germany—7.40	Italy—9.28
• Italy—2.32	Germany—7.48
• Cz. Rep—1.49	China—2.20
• USA—0.87	U S A—1.85
• China—0.52	France—1.67
• Belgium—0.42	Japan—1.29
• Spain—0.37	Belgium—0. 97
• Australia—0.32	U K—0.78
• Greece—0.15	Australia—0.77
• Slovakia—0.14	Greece—0.42

Reference: EPAI: GMO_2015/16

Solar PV—Installation growth in cumulative capacity (Giga watts):

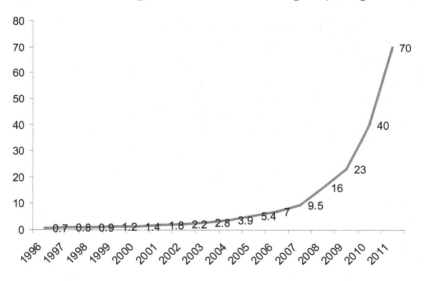

World Cumulative PV (MW) installed (region wise)—2011

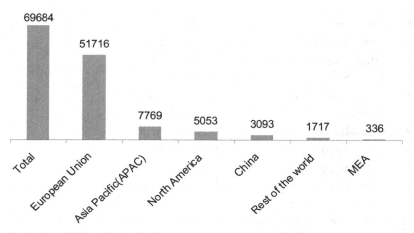

Reference: EPIA: GMO_2016

World Cumulative (2011) PV power (MW) installation by country

Country	PV installation in GW	% share in worldwide installation
Germany	24.7	36
Italy	12.8	18
Rest of world	7.0	11
Japan	4.9	7
Spain	4.4	6
USA	4.3	6
China	3	4
France	2.7	4
Belgium	2	3
Czech Repub	1.9	3
Australia	1.3	2
India	0.5	0.3

Reference: EPIA: GMO_2016

Leading solar module manufacturers (GW)

- First Solar —2—USA
- Suntech Power —1.8—China
- Yingli Green Power —1.55—China
- Trina Solar —1.4—China
- Canadian Solar —1.36—Canada
- Sharp —1.15—Japan
- Hanwha Solar One —0.82—Korea
- Jinko Solar —0.78—China
- LDK Solar —0.77—China
- Solar World —0.76—Germany

As seen above 5 companies are from China, 2 from USA, 1 each from Germany, Japan, Korea & Canada. This contributes to 40 to 50% of total global shipment. Due to falling price of modules and industry consolidation many of these giants are reworking their strategy to make their business viable.

Solar PV for rural off grid:

Solar PV for off grid applications has been in use since the eighties. The nineties saw a large number of companies and countries identifying this as an ideal option for rural households with no or a little access to electricity. Primarily catering to the rural unelectrified market, these Solar PV systems support the households for providing basic light. Dc systems are the best fit here. Systems ranging from 2 lights to 4 lights and a Dc fan are in use. The DC lights used were cfl of 5 to 11 watts. Recently good energy efficient LED lights are also in use.

Solar DC systems have provided lights to millions living at far remote areas with no proper travel or communication access. These systems replaced the traditional kerosene lamps or candles providing the households with safe and quality light. A large number of countries including South Africa, Rural India, Sri Lanka, Philippines, Indonesia, South Western provinces of China, Bangladesh, and Cambodia have extensively adopted solar DC systems. Almost all these markets have evolved around a sound financing scheme for purchase of these systems by the economically backward. World Bank has supported various initiatives in these countries directly and through intermediatories by offering credit guarantees and subsidies or both.

Solar water heating capacity as of 2010

Top 10 countries (GWth)—World Total is 195.8 Gwth

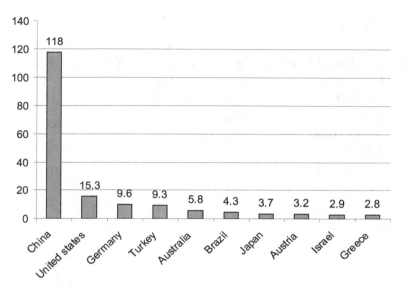

Reference: SHC 2012

Solar thermal heating market has seen tremendous growth. China is dominant with a capacity of 70% of total installed capacity world wide. With a low temperature and sunny climate, Solar water heaters became and ideal product for deployment and usage. Wide adoption of Evacuated Tube All glass technology with a lower price point also accelerated the market.

Thermal installation done in 2010—Top Ten countries

- China—34300 MWth
- Turkey—1160 MWth
- United States—814 MWth
- Germany—805 MWth
- Australia—754 MWth
- Brazil—677 MWth
- India—622 MWth

- Italy—343 MWth
- Spain—243 MWth
- France—228 MWth

Reference: SHC 2012

Evolution of world annual Thermal market (GWth)

2000—6.5 GWth   2006—16 GWth
2001—8 GWth   2007—20 GWth
2002—9 GWth   2008—27.5 GWth
2003—10.5 GWth   2009—35 GWth
2004—12 GWth   2010—42 GWth
2005—13.5 GWth

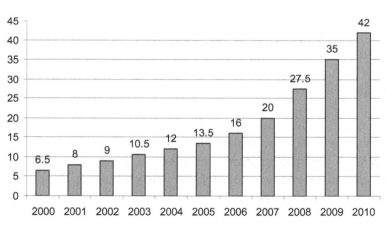

Reference—Werner wises / Franz Mauthner: SHW 2012

EU is a traditional market for Solar thermal heating and effectively used this Flat plate technology to harness the Sun for heating purpose. Turkey and Japan are early starters while United States and India are now a vibrant solar thermal market.

Countries leading in solar installations:

Overall Position 1—Germany:
Capital: Berlin
Location: Latitude—52.52° N, Longitude 13.41° E

Population: 82 mln
GDP: 3.60 trln usd
Energy Use (TOE per capita): 3.89
Co2 Emission (tons per capita) :9
Average Insolation: 1.4 to 4.4 kWh/m²/day
Solar Photovoltaic:

- Global rank 1 in installed capacity
- As of 2011 the total installed (Gw)capacity was 24.7
- The installation (GW) added in 2011 was 7.48

Solar Thermal:

- Global rank 3 in installed capacity
- The total installed capacity (GWth)of collectors is 9.6
- The total installed area (km²) of collectors is 13.75

National Renewable energy support policies:

- Feed in tariff
- Heating obligation / mandate
- Capital Subsidy
- Reduction in Vat, other taxes
- Public investment, loans or grants

Overall Position 2—USA:
Capital: Washington DC
Location: Latitude—38.89 ° N, Longitude 77.03 ° W
Population: 307 mln
GDP: 15.0 trln usd
Co2 Emission (tons per capita) :17
Energy Use (TOE per capita): 7.08
Average Insolation: 3 to 7 Kwh/ (m²-day)
Solar Photovoltaic:

- Global rank 5 in installed capacity
- As of 2011 the total installed capacity (GW) was 4.3
- The installation (GW) added in 2011 was 1.85

Solar Thermal:

- Global rank 2 in installed capacity
- The total installed capacity (GWth) of collectors is 15.3
- The total installed area (km²) of collectors is 21.88

National Renewable energy support policies:

- Capital subsidy
- Investment or production tax credit
- Reduction in Vat or other taxes
- Public investment, loans or grants

Overall Position 3—China:
Capital: Beijing
Location: Latitude—39.91 ° N, Longitude 116.4 ° E
Population: 1331.46 mln
GDP: 7.3 trln usd
CO_2 Emission (tons per capita) :6
Energy Use (TOE per capita): 1.68
Average Insolation: 2 to 6.5 Kwh/ (m²-day)
Solar Photovoltaic:

- Global rank 6 in installed capacity
- As of 2011 the total installed capacity (GW) was 3
- The installation (GW) added in 2011 was 2.2

Solar Thermal:

- Global rank 1 in installed capacity
- The total installed capacity (GWth) of collectors is 118
- The total installed area (km²) of collectors is 168

National Renewable energy support policies:

- Feed in tariff
- Capital subsidy
- Public investment, loans or grants

- Electrical utility mandate
- Heating obligation / mandate

Overall Position 4—Japan:
Capital: Tokyo
Location: Latitude—35.68 ° N, Longitude 139.76 ° E
Population: 127.56 mln
GDP: 5.90 trln usd
Co2 Emission (tons per capita) :9
Energy Use (TOE per capita): 3.71
Average Insolation: 2 to 4 Kwh/ (m²-day)
Solar Photovoltaic:

- Global rank 3 in installed capacity
- As of 2011 the total installed capacity (GW) was 4.9
- The installation (GW) added in 2011 was 1.29

Solar Thermal:

- Global rank 7 in installed capacity
- The total installed capacity (GWth) of collectors is 3.7
- The total installed area (km²) of collectors is 5.78

National Renewable energy support policies:

- Feed in tariff
- Net metering
- Capital subsidy
- Public investment, loans or grants
- Electrical utility mandate

Overall Position 5—Australia:
Capital: Canberra
Location: Latitude—35.28 ° S Longitude—149.13 ° E
Population: 22.7 mln
GDP: 1.37 trln usd
Co2 Emission (tons per capita) :18
Average Insolation: 3.75 to 6.75 Kwh/ (m²-day)

Solar Photovoltaic:

- Global rank 10 in installed capacity
- As of 2011 the total installed capacity (GW) was 1.3
- The installation (GW) added in 2011 was 0.77

Solar Thermal:

- Global rank 5 in installed capacity
- The total installed capacity (GWth) of collectors is 5.8
- The total installed area (km^2) of collectors is 8.31

National Renewable energy support policies:

- Capital Subsidy
- Public investment, loans or grants

Overall Position 6—Italy:
Capital: Rome
Location: Latitude—41.9 ° N, Longitude 12.5 ° E
Population: 61 mln
GDP: 2.20 trl usd
Energy Use (TOE per capita): 2.70
CO_2 Emission (tons per capita) :7
Average Insolation: 3.4 to 5.4 Kwh/ (m^2-day)
Solar Photovoltaic:

- Global rank 2 in installed capacity
- As of 2011 the total installed capacity(GW) was 12.8
- The installation(GW) added in 2011 was 9.28

Solar Thermal:

- The total installed capacity (GWth) of collectors is 1.8
- The total installed area (km^2) of collectors is 2.6

National Renewable energy support policies:

- Feed in tariff

- Electrical utility mandate
- Net metering
- Heating obligation / mandate
- Capital subsidy

Overall Position 7—Spain:
Capital: Madrid
Location: Latitude—40.40 ° N, Longitude 3.68 ° W
Population: 45.96 mln
GDP: 1.46 trln usd
Energy Use (TOE per capita): 2.79
CO_2 Emission (tons per capita) :6
Average Insolation: 3.3 to 5.2 Kwh/ (m²-day)
Solar Photovoltaic:

- Global rank 4 in installed capacity
- As of 2011 the total installed capacity (GW) was 4.4
- The installation (GW) added in 2011 was close to nil

Solar Thermal:

- The total installed capacity (GWth) of collectors is 1.7
- The total installed area (km²) of collectors is 2.46

National Renewable energy support policies:

- Feed in tariff
- Investment or production tax credit
- Reduction in Vat or other taxes
- Public investment, loans or grants
- Heating obligation / mandate

Overall Position 8—France:
Capital: Paris
Location: Latitude—48.87 ° N, Longitude 2.34 ° E
Population: 62.62 mln
GDP: 2.77 trln usd
Energy Use (TOE per capita): 4.04
CO_2 Emission (tons per capita) :6

Average Insolation: 2.7 to 5.2 Kwh/ (m²-day)
Solar Photovoltaic:

- Global rank 7 in installed capacity
- As of 2011 the total installed capacity (GW) was 2.7
- The installation (GW) added in 2011 was 1.67

Solar Thermal:

- The total installed capacity (GWth) of collectors is 1.5
- The total installed area (km²) of collectors is 2.28

National Renewable energy support policies:

- Feed in tariff
- Capital subsidy
- Investment or production tax credit
- Reduction in Vat or other taxes
- Public investment, loans or grants

Overall Position 9—Turkey:
Capital: Ankara
Location: Latitude—39.91 ° N Longitude—32.85 ° E
Population: 74.8 mln
GDP: 0.770 trln USD
Energy Use (TOE per capita): 1.44
CO_2 Emission (tons per capita) :4
Average Insolation: 3.6 Kwh/ (m²-day)
Solar Photovoltaic:

- As of 2010 the total installed capacity (MW) was 3
- The installation (MW) added in 2010 was 1

Solar Thermal:

- Global rank 4 in installed capacity
- The total installed capacity (GWth) of collectors is 9.3
- The total installed area (km²) of collectors is 13.3

National Renewable energy support policies:

- Feed in tariff

Overall Position 10—Greece:
Capital: Athens
Location: Latitude—37.97 ° N, Longitude 23.73 ° E
Population: 11.29 mln
GDP: 0.30 trln usd
Energy Use (TOE per capita): 2.57
CO_2 Emission (tons per capita) :8.4
Average Insolation: 3.8 to 5.2 Kwh/ (m²-day)
Solar Photovoltaic:

- As of 2011 the total installed capacity (GW) was 0.63
- The installation (GW) added in 2010 was 0.42

Solar Thermal:

- Global rank 10 in installed capacity
- The total installed capacity (GWth)of collectors is 2.81
- The total installed area (km²) of collectors is 4.1

National Renewable energy support policies:

- Feed in tariff
- Net metering
- Capital subsidy
- Investment or production tax credit

Overall Position 11—India:
Capital: New Delhi
Location: Latitude—28.67 ° N, Longitude 77.21 ° E
Population: 1240 mln
GDP: 1.84 trln usd
Energy Use (TOE per capita): 0.56
CO_2 Emission (tons per capita) :2
Average Insolation: 4 to 6.5 Kwh/ (m²-day)

Solar Photovoltaic:

- Global rank 15 in installed capacity
- As of 2011 the total installed capacity (GW) was 0.5
- The installation (GW) added in 2011 was 0.3

Solar Thermal:

- Global rank 12 in installed capacity
- The total installed capacity (GWth) of collectors is 2.7
- The total installed area (km²) of collectors is 3.98

National Renewable energy support policies:

- Feed in tariff
- Net metering
- Capital subsidy
- Public investment, loans or grants
- Electrical utility mandate

Overall Position 12—Belgium:
Capital: Brussels
Location: Latitude—50.84 ° N, Longitude 4.36 ° E
Population: 10.79 mln
GDP: 0.52 trln usd
Energy Use (TOE per capita): 5.17
CO_2 Emission (tons per capita) : 10
Average Insolation: 2.9 Kwh/ (m²-day)
Solar Photovoltaic:

- Global rank 8 in installed capacity
- As of 2011 the total installed capacity (GW) was 2
- The installation (GW) added in 2011 was 0.97

Solar Thermal:

- The total installed capacity (GWth) of collectors is 0.25
- The total installed area (km²) of collectors is 0.36

National Renewable energy support policies:

- Net metering
- Capital subsidy
- Investment or production tax credit
- Reduction in Vat or other taxes

Overall Position 13—Brazil:
Capital: Brasilia
Location: Latitude—15.47 ° S Longitude—47.91 ° W
Population: 194 mln
GDP: 2.48 trln usd
Energy Use (TOE per capita): 1.24
CO_2 Emission (tons per capita) :1.9
Average Insolation: 4.5 to 6 Kwh/ (m²-day)
Solar Photovoltaic:

- As of 2010 the total installed capacity was close to nil
- The installation added in 2010 was closed to nil

Solar Thermal:

- Global rank 6 in installed capacity
- The total installed capacity (GWth) of collectors is 4.3
- The total installed area (km²) of collectors is 6.11

National Renewable energy support policies:

- Reduction in Vat or other taxes
- Public investment, loans or grants

Overall Position 14—Cezh Republic:
Capital: Pargue
Location: Latitude—50.08 ° N Longitude 14.41 ° E
Population: 10.49 mln
GDP: 0.20 trln usd
Energy Use (TOE per capita): 4.14
CO_2 Emission (tons per capita) :10
Average Insolation: 2.5 to 3.1 Kwh/ (m²-day)

Solar Photovoltaic:

- Global rank 9 in installed capacity
- As of 2011 the total installed capacity (GW) was 1.9
- The installation (GW) added in 2011 was close to nil

Solar Thermal:

- The total installed capacity (GWth) of collectors is 0.3
- The total installed area (km²) of collectors is 0.45

National Renewable energy support policies:

- Feed in tariff
- Capital subsidy
- Investment or production tax credit
- Reduction in Vat or other taxes

Overall Position 15—Austria:
Capital: Vienna
Location: Latitude—48.20 ° N Longitude—16.37 ° E
Population: 8.36 mln
GDP: 0.42 trln usd
Energy Use (TOE per capita): 3.90
Co2 Emission (tons per capita) :7
Average Insolation: 2.8 Kwh/ (m²-day)
Solar Photovoltaic:

- As of 2010 the total installed capacity (MW) was 103
- The installation (MW) added in 2010 was 50

Solar Thermal:

- Global rank 8 in installed capacity
- The total installed capacity (GWth) of collectors is 3.2
- The total installed area (km²) of collectors is 4.55

National Renewable energy support policies:

- Feed in tariff
- Capital subsidy
- Investment or production tax credit
- Public investment, loans or grants

Overall Position 16—Israel:
Capital: Jerusalem
Location: Latitude—31.78 ° N Longitude—35.21 ° E
Population: 7.4 mln
GDP: 0.240 trln usd
Energy Use (TOE per capita): 2.87
CO_2 Emission (tons per capita) :9
Average Insolation: 5.48 Kwh/ (m²-day)
Solar Photovoltaic:

- As of 2010 the total installed capacity was close to nil
- The installation added in 2010 was close to nil

Solar Thermal:

- Global rank 10 in installed capacity
- The total installed capacity (GWth) of collectors is 2.9
- The total installed area (km²) of collectors is 4.2

National Renewable energy support policies:

- Feed in tariff
- Reduction in Vat or other taxes
- Heating obligation / mandate

—The above listed thermal figures are for year 2010
Reference: Solar Heat Worldwide_2012Edition _IEA
Global Market outlook until 2016—EPIA
REN21_GSR_2012/11
World Bank

Renewable energy roadmap around the world:

Many countries around the world have now about 5% share of the final energy now coming from renewable energy. This will be increased to a minimum of 15% and in some countries going up to 50% by 2020.

Serial No	Country	Renewable energy 2009 share—% of Final energy	Renewable energy Future target—% of Final energy	Target Year
1	Austria	29	34	2020
2	Belgium	3.8	13	2020
3	China	9.1	15	2020
4	Czech Republic	8.5	13	2020
5	Denmark	20	30	2020
6	France	12	23	2020
7	Gabon		80	2020
8	Germany	9.7	60	2050
9	Greece	7.9	18	2020
10	Israel		50	2020
11	Italy	7.8	17	2020
12	Latvia	37	40	2020
13	Madagascar		54	2020
14	Netherlands	4.2	14	2020
15	Palestinian Territories		20	2020
16	Portugal	26	31	2020
17	Romania	22	24	2020
18	Slovenia	18	25	2020
19	Spain	13	20	2020
20	United Kingdom	2.9	15	2020

Reference: REN21_GSR_2011

Jobs in renewable energy

An estimated 1.5million people are now employed in solar industry Worldwide, jobs in renewable energy industries exceeded 3.5 million in 2010. A 2008 report on jobs in renewable energy observes that while developed economies have shown the most technological leadership in renewable energy, developing countries are playing a growing role and this is reflected in employment.

China, Brazil, and India account for a large share of global total employment in renewables, having strong roles in the wind power, solar hot water, and/ or biofuels industries. In addition to manufacturing, many of these jobs are in installations, operations, and maintenance, as well as in biofuels feedstocks. Jobs are expected to grow apace with industry and market growth, although increasing automation of manufacturing and economies of scale in installation services may moderate the rate of jobs growth below that of market growth.

Jobs associated with the on-shore wind industry are 15 person-years in construction and manufacturing per MW produced, and 0.4 jobs in operations and maintenance per MW existing, according to the European Wind Energy Association (2009). Similar estimates for the solar PV sector are 38 person-years per MW produced and 0.4 jobs per MW existing, according to the European Photovoltaics Industry Association.

Future growth:

The solar energy has shown very impressive growth in the last decade. The solar PV market has grown at a rate near to 33% CAGR from 2000 to 2005 and then doubled the growth rate at about 60% CAGR from 2006 to 2010. The second half can be seen as an explosive growth supported by Germany, Spain and Italy. This decade will see Japan and China aggressively increasing their contribution while Germany will continue to lead.

Considering the current state of world economy and the consolidation happening in the solar worldwide the CAGR here is taken in a more realistic way, there by seen moving towards 30% by the next two decades similar to a high growth mature industry.

The solar Thermal market consisting of low and medium heat (40 deg to 90 deg centigrade) has grown at a rate near to 11% CAGR from 2000 to 2005 and then almost doubled the growth rate at about 21% CAGR from 2006 to 2010. The second half can be seen as an explosive growth supported by Asian countries

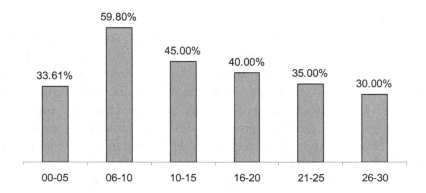

especially China. This decade will see Turkey, United States and India aggressively increasing their contribution while China will continue to increase its lead. The growth rate is expected to

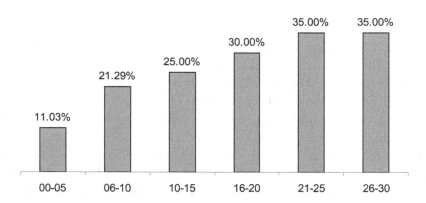

steadily increase to 30% during this decade and stabilize at 35% by 2030.

Considering the above growth rate the market is expected to surpass the 100GW figure mark by 2015 and

Annual Worldwide Market for Solar—PV & Thermal

Year	2010	2015	2020	2025	2030
Solar PV(GW)	16	100	500	2500	10000
Solar Thermal (GWth)	42	125	475	2100	10000

touch 10,000 GW by year 2030. The total installed capacity of Solar PV is expected to reach 1000GW by 2018, 5000 by 2023 and cross 30000GW by 2030. Similarly the total installed capacity of solar thermal will meet the 1000GWth by 2017, 5000GWth by 2023 and touch 30000GWth by 2030.

The CSP (Concentrating Solar Power) is a very promising concept which is showing good signs of growth. This very high temperature solar thermal category has been looked upon for the last 10 to 15 years, but now has become serious business even though the cost is high. A number of power plants is in place and new are being build. The total installed capacity now is over 1 GW and expected to reach 10 GW by 2015.

On the PV technology front the thin film will gain more prominence with larger semiconductor manufacturers from the east joining the industry to drive the cost further down, higher efficiency inverters will get more attention and battery materials will move from lead to lithium or such to make compact and versatile batteries. On the solar thermal front with the copper price increasing steadily, aluminum will be increasingly used as absorber material with evacuated glass tubes continuing to capture major share.

The regions of growth for Solar PV in the short term (less than 5 years) will continue to be Europe (powered by Germany), USA and Japan. In the longer term the Asian giants China and India will establish themselves as world market leaders. For Solar Thermal in the short term (less than 5 years) growth will continue to be Asia (powered by China), Germany and United States. In the longer term the Turkey, Australia, India etc will join the above to propel the market.

China's Golden Sun Program and India's JNSM along with the renewable energy roadmap of different countries seen above will be the pillars of future growth. Significant investment in research and development, design, education and training is expected to happen as the industry matures. A number of doctoral and graduate programs have sprung up around the world in this respect. There will be a stage of consolidation and re growth in the medium term.

The continued aggressiveness of the industry, confidence of the customers and commitment from the administrators/ national leaders will be the key enablers for this extreme growth scenario to maintain its path.

CHAPTER 7

INDIA ELECTRIC POWER SCENARIO

"A nation's strength ultimately consists in what it can do on its own and not in what it can borrow from others."
—*Smt Indira Gandhi*

India Electric Energy Dash Board:

Installed capacity (Apr'12)	Power supply position (Apr '12)	Power demand position (Apr '12)	Per capita electricity consmptn (10-11)	No of Household connections
202 GW	Deficit Over 8 Billion Units	Deficit Over 12 Giga Watts	550 KWh (approx)	Over 16 crores
Electricity consmptn (2010-11)	T & D losses (09-10)	AVG tariff of electricity -2010 (Domestic)	Peak tariff of electricity -2010 (Commercial)	Primary Energy consmptn (10-11)
694 TWhr	20%	Rs 3.54/ unit	Rs 12/ unit	494 MTOE

Reference: Energy statistics 2012
CEA Publications

Sources of Power in India

1) Coal (2) Diesel (3) Natural Gas (From Fossil Fuel)

In a fossil fuel power plant the chemical energy stored in fossil fuels such as coal, fuel oil, natural gas and oxygen of the air is converted into thermal energy to generate mechanical power and, then to electrical power for continuous use and distribution across a wide geographic area. Multiple generating units may be built at a single site for more efficient use of land, natural resources and labour.

Coal Power plants form bulk of the energy generation in India. There are 120 odd coal power plants. The bigger ones are

Coal Power Plants	Design Capacity (Mwe)	State
Tata Mundra Coal Power Plant	4000	Gujarat
Vindhyanchal STPS Coal Power Station	3260	Madhya Pradesh
Talcher Kaniha STPS Coal Power Station	3000	Orrisa
Kahalgoan STPS Coal power Station	2340	Bihar
Chandrapur Coal Power Station	2340	Maharashtra

There are about 8 Diesel power plants in India. The capacity range from 158 Mwe to 20 Mwe. Mostly these are situated in the southern states of Karnataka, Tamil Nadu & Kerala. The largest of them is the Diesel power plant at Whitefield Industrial Park—Karnataka with a design capacity of 158 Mwe.

The total number of Natural gas power plants in India is about 62.
The design capacity range from 1100Mwe to 20 Mwe.

Natural gas Power Station	Design Capacity (Mwe)	State
SUGEN(Torrent) CCGT power plant	1145	Gujarat
Dadri Gas CCGT power plant	817	Uttar Pradesh
Uran Gas CCGT power plant	912	Maharshtra
Rathnagiri 1, 2, &3 Gas CCGT power plant	2220	Maharshtra

4) Nuclear

The basic operation can be summarized as—The heat is produced by fission in a nuclear reactor (corresponding to a boiler in a coal power plant) and given to a heat transfer fluid—usually water (for a water reactor). Directly or indirectly water vapor-steam is produced. The pressurized steam is then usually fed to a multi-stage steam turbine. After the steam turbine has expanded and partially condensed the steam, the remaining vapor is condensed in a condenser. The condenser is a heat exchanger which is connected to secondary side such as a river or a cooling tower. The water then pumped back into the nuclear reactor and the cycle begins again.

There are about 6 Nuclear power plants in India with a design capacity ranging from 1400Mwe to 440 Mwe. The larger ones being

Nuclear Power Plant	Design Capacity (Mwe)	State
Tarapur Atomic power station	1400	Maharshtra
Rajasthan Atomic Power Station	1180	Rajasthan

5) Hydro Power

Hydroelectricity is generated through the use of the gravitational force of falling or flowing water. It is the most widely used form of renewable energy. A hydroelectric complex produces no direct waste, and has a considerably lower output level of the greenhouse gas carbon dioxide (CO_2) than fossil fuel powered energy plants.

Hydroelectric power plants are the highest in number in the category of power plants. The total number of hydro electric power plants in India is about 140. But unlike the Coal power plants these are smaller in generation capacity. The design capacity range from 2000 Mwe to 4 Mwe. A few of the prominent hydro electric power plants are

Hydro Electric power plants	Design Capacity (Mwe)	State
Tehri Hydro electric power project	2000	Uttarakand
Koyna Hydro electric power plant	1920	Maharshtra
Naptha Jhakri hydro electric power plant	1500	Himachal pradesh

6) Renewables

Wind Power, Bio mass & Solar power form the renewable energy power station. India has about 18 wind farms. India has the 4th largest wind installation in the World. Bio mass and Solar power plants are a new foray for India. A few of these are established and many are upcoming.

Electric Power—Current status:

India is the world's sixth largest energy consumer, relying on coal as the primary energy source for over half of its total energy needs.

Fuel wise Installed capacity—2012(April) in MW

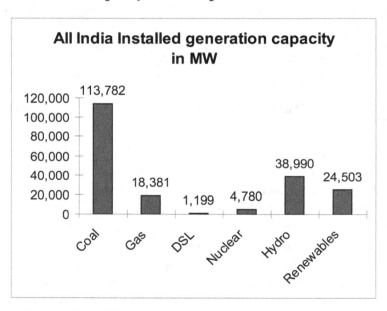

Reference: CEA Publications

In GW

Region	Thermal				Nlr	Hdr	Rnl	Total
	Cl	Gas	DSL	Total				
North	29	4.4	0.01	33.5	1.6	15.1	4.4	54.5
West	39.4	8,2	0.02	47.7	1.8	7.5	8	65

South	23	4.7	0.9	28.5	1.3	11.3	11.5	52.7
East	22.4	0.2	0.02	22.6	0	3.9	0.39	26.8
N E	0.06	0.8	0.14	1	0	1.12	0.22	2.4
Islands	0	0	0.07	0.07	0	0	0.006	0.076
All India	114	18	1.2	133	4.8	39	24.5	201.6

Cl—Coal; DSL—Diesel: Nlr—Nuclear: Hdr—Hydro;
Rnl—Renewables
Reference: CEA Publications

Thermal power plants produce more than three quarters of India's electricity, taking advantage of India's position as the third largest producer of coal in the world.

The electricity sector has long experienced capacity shortfalls, poor reliability and quality of electricity (voltage fluctuation, etc.) and frequent blackouts. Industry cites electricity supply as a major impediment to economic growth.

Power Supply Status—Units (2011-12)

Region	Power Supply Requirement (Million Units)	Power Supply Availability (Million Units)
All India	936568	857239

	Power Supply Requirement (Million Units)	Power Supply Availability (Million Units)
Northern Region	275820	258140
Western region	290423	257408
Southern Region	259934	237113
Eastern Region	99380	94614
North Eastern Region	11011	9964

Region Wise Power Supply position 11-12

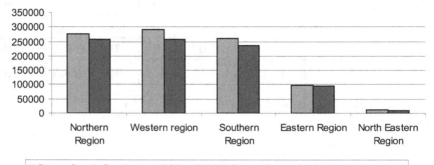

Reference: CEA Publications

Power Supply status—Demand

Region	Peak Demand (MW)	Demand Met (MW)
All India	130250	115847

Region	Peak Demand (MW)	Demand Met (MW)
Northern Region	40248	37117
Western region	42352	36509
Southern Region	38121	32188
Eastern Region	14505	13971
North Eastern Region	1920	1782

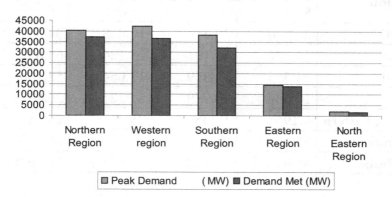

Reference: CEA Publications

Both in terms of units consumed / required and demand, the current installed capacity is lagging behind creating huge deficits. Western region tops this list of highest shortage of electrical power.

Power generating Installed capacity—Ownership Wise in MW _ April 2012

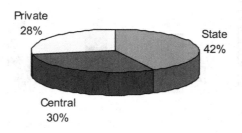

State	Central	Private
85,916	60,180	55,533

Reference: CEA Publications

Region	Ownership			Total (MW)
	State	Central	Private	
North	24,040	20,456	10,088	54,584
West	25,577	16,191	23,124	64,892
South	24,365	11,319	17,054	52,738
East	10,684	10,979	5,222	26,885
North Eastern	1,195	1,235	25	2,455
Islands	55	0	21	76
All India	85,916	60,180	55,533	201,630

Despite reforms introducing private participation during the 1990s, the India's electricity sector has remained dominated by the state since India's independence in 1947. The Electric Supply Act of 1948 integrated smaller fragmented utilities into 19 state electricity boards. SEBs remain

Top 5 states in terms of Installed capacity

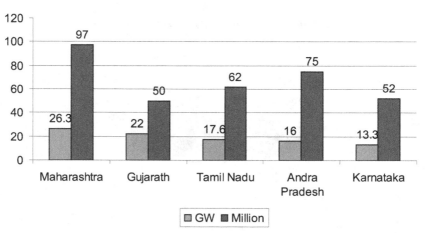

Reference: CEA Publications

State	Installed capacity(GW)	Population Million
Maharashtra	26.3	97
Gujarath	22	50
Tamil Nadu	17.6	62
Andra Pradesh	16	75
Karnataka	13.3	52

the dominant institutions within India's electricity industry, controlling well over half of the electricity supply and the vast majority of distribution.

The Indian Constitution lists electricity as a "concurrent" responsibility of the state and central governments, meaning that the state legislature's authority overlaps with the central government. In the event of a conflict between overlapping state and federal authority, the parliament in New Delhi can exercise preemptive power.

India faces formidable challenges in meeting its energy needs and in providing adequate energy of desired quality in various forms in a sustainable manner and at competitive prices. India needs to sustain an 8% to 10% economic growth rate, over the next 25 years, if it is to eradicate poverty and meet its human development goals.

Growth in installed capacity

Year	1985	1990	1992	1997	2002	2003	2004
GW	43	64	69	85	105	107	112
Year	2005	2006	2007	2008	2009	2010	2012
GW	118	124	132	143	148	160	201

Growth in Installed capacity

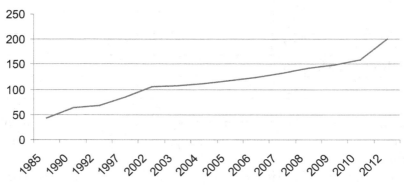

Reference: CEA Publications

All India Electricity per Capita Consumption

Year	Per capita electricity consumption (Kwh)
1970-71	79
1980-81	120
1990-91	223
2000-01	306
2005-06	369
2008-09	478
2009-10	521

To deliver a sustained growth rate of 8% through 2031-32 and to meet the lifeline energy needs of all citizens, India needs, at the very least, to increase its primary energy supply by 3 to 4 times and, its electricity generation capacity/supply by 5 to 6 times.

Population without electricity

Country	India	China	Indonesia	Pakistan	South Africa	Brazil	Iran
Population in million	404	8	81	70	12	4.3	1.2

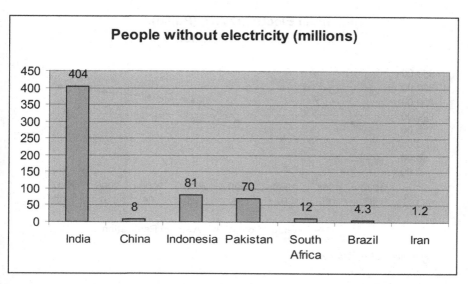

Source: WEO_IEA Publications

India Electricity Cost Avg:

Cost of supply vs. Realisation (Paise)

Year	Average Cost	Average Realisation
1980-81	41.00	32.30
1984-85	65.07	49.39
1990-91	112.32	86.84
1995-96	179.38	151.28
2000-01	246.00	181.00
2005-06	260.00	221.00
2006-07	276.00	227.00
2007-08	293.00	239.00

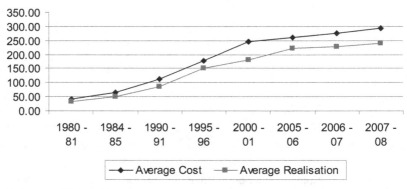

Reference: CEA Publications

Commercial losses of power utilities in India:

The SEBs fall under the jurisdiction of individual state governments. Currently, the financial losses of the SEBs are mounting and it is over 1% of India's GDP. The SEBs which are the engines of power generation and distribution are often subjected to political decisions / indecisions rather than those with logic and economic sense. This is resulting in the bleeding of these important institutions of our country. A customer centric fast paced SEBs with little political intervention can utilize the need requirement situation to improve their business and reduce the losses. The electric power being one of the most sought out services throughout the country only supports this.

Year	Commercial Loss (Cr in Rs)
1990-91	3000
1994-95	6125
1999-00	26353
2001-02	29331
2003-04	19107
2005-06	20869
2007-08	31862

The previous section showed the widening gap between cost of supply and realization. Even though average consumers pay Rs 3.5

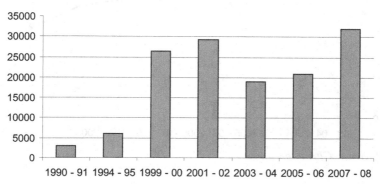

Reference: CEA Publications

for one unit of energy used, the average cost realization of power utilities is only about Rs 2.5 per unit. Since consumers pay almost equal of the cost of production, the T&D and other losses results in reduced average realization.

With growth in electricity consumption, the trend shows increase in commercial loss of the power utilities. This need to be arrested or else the nation's power providers will soon be in dire straits.

All India Transmission & Distribution Losses:

Year	2003-04	2004-05	2005-06	2006-07	2007-08	2008-09	2009-10
T&D loss in %	32.53	31.25	30.42	28.65	27.2	25.47	20

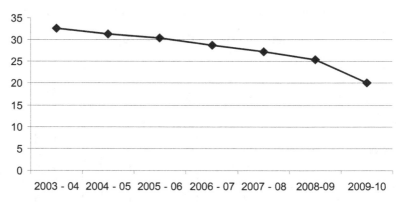

Reference: CEA Publications

State Wise Transmission & Distribution losses

Serial No	State / Union territory	Loss in % (2006-07)
1	Haryana	33.45
2	Himachal Pradesh	19.77
3	Jammu & Kashmir	51.98
4	Punjab	26.61
5	Rajasthan	35.60
6	Uttar Pradesh	33.49
7	Uttarakhand	34.48
8	Chandigarh	25.13
9	Delhi	33.00
10	Gujarath	24.87
11	Madhyapradesh	39.24
12	Goa	20.90
13	Daman & Diu	22.09
14	Andhra Pradesh	18.65
15	Karnataka	25.91
16	Kerala	19.11
17	Tamilnadu	19.54
18	Lakshadweep	12.87

19	Puducherry	18.76
20	Bihar	50.67
21	Jharkhand	26.21
22	Orissa	40.86
23	Sikkim	26.86
24	West Bengal	23.86
25	Assam	33.69
26	Manipur	53.47
27	Meghalaya	35.34
28	Nagaland	54.79
29	Tripura	34.75
30	Arunachal Pradesh	57.79
31	Mizoram	38.18

The above published figures indicate that the loss in transmission across the states. For a country that is reeling under acute power shortages, the losses in distribution will add to the misery. The distribution losses are seen over 50% in some states which need immediate attention. The national average T & D losses are now brought down lower than 20% by official estimate. Still this is on a very large scale. Hence community based power generation should be encouraged where ever possible which will reduce these losses.

Power Supply to Agriculture (Sept_2010)

State / Union Territory	Average hours of supply
Chandigarh	24 hrs
Delhi	24 hrs
Haryana	3 phase: 9.5 hrs
Himachal Pradesh	24 hrs
Jammu & Kashmir	NA
Punjab	3 phase : 11 hrs
Rajasthan	3 phase : 4 hrs
Uttar Pradesh	3 phase : 10.44 hrs

Uttarakhand	3 phase : 22 hrs
Chattisgarh	3 phase : 18 hrs
Gujarath	8 hrs
Madhya Pradesh	3 phase : 18 hrs
Maharshtra	3 phase : 17 hrs
Goa	No restriction
Andhra Pradesh	3 phase : 7 hrs
Karnataka	3 / 1 phase : 6 hrs
Kerala	No restriction
Tamil nadu	3 phase: 9 hrs ; 1 phase: 15 hrs
Puducherry	No restriction
Bihar	18 hrs
Jharkhand	20 hrs
Orissa	24 hrs
West bengal	24 hrs

Agriculture, the main stay of India's economic growth receives only 1/3 of the electricity required now. The above table shows restricted supply in almost all the states. This will severely affect the water pumping and allied support to agriculture. We see very large tracts of land uncultivated due to water scarcity and flooding. This can be put to agricultural use only if adequate power is provided. The central and state governments need to enhance its support to agriculture and ensure uninterrupted power supply for a robust growth in agricultural production. This power need to come from either central or state grid or new sources where solar can have a good role to play.

Sector mix of electricity consumption:

Total consumption = 694392 (2010-11) (All in Gwhr)

Industry	Agriculture	Domestic	Commercial	Traction	others
272589	131967	169326	67289	14003	39218
39.26%	19.00%	24.38%	9.69%	2.02%	5.65%

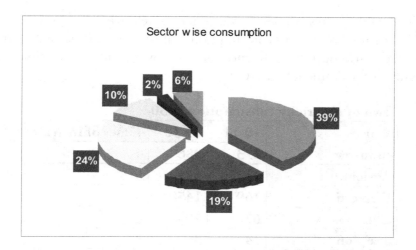

Here we see the highest consumption comes from the Industry followed by Domestic and then Agriculture.

The growth rate for the last two years in electricity for Industry, Agriculture and Domestic were 15 %, 10 % and 16 % respectively while the total consumption grew by about 13 %. On the CAGR 1970 to 2010 also the consumption of domestic was highest at 10% while industry and agriculture where 6% and 8.5% respectively and others at a lower level. The consumption of electricity for domestic purpose has a significant impact on the total consumption.

Future requirement of electricity:

Considering that the current economic growth will continue, the final consumption of electricity will increase many fold from the present requirements by 2029-30, growing at CAGR of 8.5%.

We look at the current annual growth rate by sectors
Industry consumption—7%
Agricultural consumption—9%
Domestic consumption—12%
Commercial—8%
Traction—5%
Others—7%

The table below shows a projection of electricity requirement in 2030. The total electricity consumption will increase 5 times with strong domestic consumption of 9 times followed by agriculture at 5times and Industry and Commercial at 4 times.

Twh of electricity consumption 2030

Year	10-11	29-30	Times of increase
Industry	273	986	4
Agriculture	132	679	5
Domestic	169	1458	9
Commercial	67	290	4
Traction	14	35	3
others	39	142	4
Total all sectors	694	3590	5

The above figures are arrived at by considering the current trend. Hence this is a conservative estimate. Actual requirement will be higher.

Industry and infrastructure requirement is expected to grow at a much higher pace. The increase in per capita consumption which is currently lower than developed and many developing countries will propel the domestic demand. Added to this is the increase in access to electricity. The consumption of domestic will exceed the Industry by year 2022 and will top the list in electricity consumption.

India Final Electricity consumption (TWh)

Year	2010	2015	2020	2025	2030
Twh	610	970	1500	2300	3500

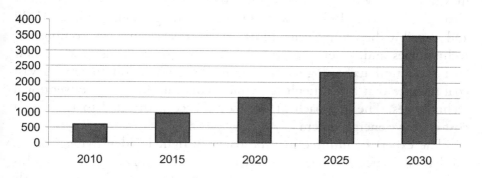

The continued modernization of agriculture will increase the energy requirement. The increase in services sector with 10% growth will increase the requirement of electricity in commercial to unprecedented level seen so far. Similarly with lay out of overhead and underground Metro trains, Mono rail and expansion of the current railway electric line will increase the energy requirement in the traction sector. In relation to all of the above the others category will need more electricity.

This means India's installed capacity should increase to at least 800 GW from the current 202 GW which in turn will take the primary energy requirement to 1.0 BTOE which is huge on the current scale.

Integrated Energy Policy—Expert committee report extract:

"Meeting the energy challenge is of fundamental importance to India's economic growth imperatives and its efforts to raise its level of human development. The broad vision behind the energy policy is to reliably meet the demand for energy services of all sectors at competitive prices. Lifeline energy needs of all households must be met even if that entails directed subsidies to vulnerable households. The demand must be met through safe, clean and convenient forms of energy at the least cost in a technically efficient, economically viable and environmentally sustainable manner.

Considering the shocks and disruptions that can be reasonably expected, assured supply of such energy and technologies at all times is essential to providing energy security for all. Meeting this vision requires that India pursues all available fuel options and forms of energy, both conventional and non-conventional. Further, India must seek to expand its energy resource base and seek new and emerging energy sources.

Finally, and most importantly, India must pursue technologies that maximize energy efficiency, demand side management and conservation. Coal shall remain India's most important energy source till 2031-32 and possibly beyond. Thus, India must seek clean coal combustion technologies and, given the growing demand for coal, also pursue new coal extraction technologies such as in-situ gasification to tap its vast coal reserves that are difficult to extract economically using conventional technologies. The approach of the Committee is directed to realising a cost-effective energy system."

The expert committee may be partially right in identifying Coal as the largest source of power for India during the coming years. But this will have to substantially change if our growth has to be for a sustainable living. Especially the domestic sector should move away from using electricity produced by fossil fuels.

CHAPTER 8

SOLAR POWER IN INDIA

> *India is deeply committed to the goals of sustainable development. We have one of the most active renewable energy programmes in the world. We have vigorously promoted the use of solar energy in both thermal and electricity generation modes. India's 5000 year old culture enjoins us to look at the whole world and all it sustains—living and non living—as a family, coexisting in a symbiotic manner*
>
> ~ *A.B. Vajpayee Prime minister of India at UNFCC_2002*

India's economic growth is now about 6 to 7%. Growth and modernization essentially follow the energy. For each 1% of economic growth, India needs around 0.75% of additional energy. India is facing a formidable challenge to build up its energy infrastructure fast enough to keep pace with economic and social changes. Energy requirements have risen sharply in recent years, and this trend is likely to continue in the foreseeable future. For GDP annual growth of 8%, the Planning Commission estimates that the commercial energy supply would have to increase at the very least by three to four times by 2031-2032 and the electricity generation capacity by five to six times over 2003-2004 levels. In 2031-2032, India will require approximately One billion tons of oil equivalent to cover its total commercial energy needs.

Consumption of energy

India's power market is confronted with major challenges regarding the quantity as well as the quality of the electricity supply. The base-load capacity will probably need to exceed 400 GW by 2017. In order to match this requirement, India must more than double its total installed capacity, which as of 2012 is 201 GW. Moreover, India's power sector must ensure a stable supply of fuels from indigenous and imported energy sources, provide power to millions of new customers, and provide cheap power for development purposes, all while reducing emissions. On the quality side, the electricity grid shows high voltage fluctuations and power outages in almost all parts of the country on many days for several hours. According to the "Global Competitiveness Report," in 2009-2010 (weighted average), India ranked 110 among 139 countries in the category "Quality of Electricity Supply." India's average power consumption has been increasing steadily and yet the quality of power is lagging far behind due to supply shortage and distribution constraints. Between 1980 and 2009, energy consumption increased by almost seven times from 85, TWh to 550 TWh, which corresponds to an average annual growth rate of approximately 7.1%. The strongest increase was the consumption by private households, which

Growth in Renewable energy (MNRE) for the last 2 decades

Year	85	90	97	02	04	05	06	08	09	10	12
GW	0	18	0.9	1.6	2.5	3.8	6.2	11	13.3	15.5	24.5

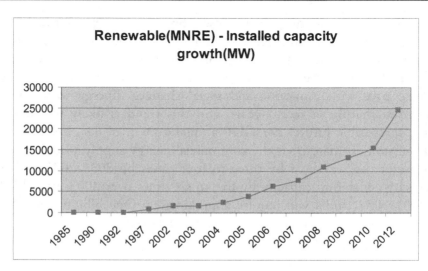

increased by almost 14 times since 1980 at an average annual growth rate of 10%. The reason for this increase was the inclusion of several million new households and corresponding increase in electrical household appliances such as refrigerators, washing machines, air conditioners, water heaters, electric cookers etc.

India due to its geo-physical location receives solar energy equivalent to nearly 5,000 trillion kWh/year, which is far more than the total energy consumption of the country today. But India produces a very negligible amount of solar energy—a mere 0.2 percent compared to other energy resources. Power generation from solar thermal energy is still in the experimental stages in India. Up till now, India's energy base has been more on conventional energy like coal and oil. However, India has aggressively charted out action plan to increase the deployment of renewable energy sources through out the country.

Solar irradiation in India:

India located in the equatorial Sun Belt receives abundant solar energy. The daily average incident varies from 4-7 kWh/m² depending on the location. The annual average radiation incident over India is about 5.5kWh/m² per day. The country experience about 300 clear sunny days every year.

The highest radiation is received in Rajasthan, northern part of Gujarat and parts of Ladakh. The parts of Andhra Pradesh, Maharashtra & Madhya Pradesh also receives very good amount of solar radiation. Any solar irradiance map shows good spread of solar irradiance throughout the country. With such high irradiance, studies have indicated that if half percent of India's available land area is brought under Solar PV, it would meet the country's energy demand in full till 2050.

Solar Energy Scenario:

A look at the Installed capacity of power generation according to contribution from energy sources is interesting. The figure below shows 10% of power generation comes from renewables. This is a decent figure even though much needed to be

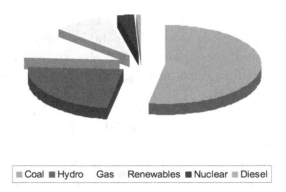

■ Coal ■ Hydro Gas Renewables ■ Nuclear ■ Diesel

done. When we look in renewable, the solar is miniscule at < 0.2% but now growing fast.

Solar products category wise deployment:

Program/ Systems	Unit	Cumulative achievement as of end 2011	Cumulative achievement (MW) (Estimate Watt peak/system)
Grid interactive Solar PV	MW	481	481
Off grid Solar PV power plants > 1 Kwp/ < 1 Kwp	MW	81/8.2	81/8.2
Solar—Wind Hybrid systems	Nos	2800	2.8 (1Kwp)
Solar Street Lightings	Nos	1,41,280	7.6 (50 Wp)
Solar Home lighting systems	Nos	6,96,069	20 (30Wp)
Solar Lanterns	Nos	8,50,375	4.25 (5Wp)
Solar PV water pumps	Nos	7,571	7.5 (1Kwp)
Total Solar PV installation (MW)			**612**
Solar Cookers	Nos	0.65 million	0.65 million
Solar water heating systems	Million—m^2	4.98	4.98

Reference: MNRE Annual Report—2011/12 and own research

The total number of actual installations will be 10 to 20% higher as installations done without the support of MNRE may not get reflected in the above table.

Solar Hot Water Installation Segmentation

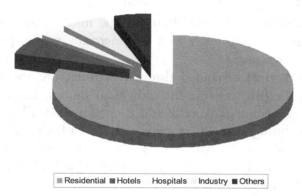

Reference: MNRE Publications

80% of solar hot water installations are in the households with hotels and industry having about 6% of the total installations.

Industry Overview:
When the oil crisis gave a thrust for renewables all over the world, here in India the renewable energy industry got under way on a serious note. Since the early stages were more technology driven and high cost involved, the initial projects were undertaken with the support of the state and premier technical institutions. A good number of academic and industrial research projects were initiated. While the Solar PV maintained a sophisticated profile limited to the R &D, the solar thermal slowly started to appear in a scattered manner.

During the mid and late eighties the solar industry started to gain momentum in both solar PV and solar thermal. Quite a few companies were formed during this period. Pilot projects for small PV systems and large thermal projects started during this time. The nineties saw the gradual spread and acceptance of solar solution with more and more customers using solar systems in their homes. This was a period of technology transfer and absorption. The state also supported the industry through various initiates. The National Aeronautical laboratories (NAL)—Bangalore, developed and commercialized Nalsun coating which is now widely used in flat plate collectors. A large number of projects for unelectrified areas were given thrust with solar PV DC systems and a few large projects. The early talks of financing through banks were initiated with participation of World Bank, UN, RBI, nationalized and grameen banks, leading solar companies, NGOs etc to support the unelectrified/ poorly electrified areas in the different parts of the country.

The late nineties and early 2000 saw the industry showing signs of adolescence and the possibility of a commercially viable, potential and mature business was seen.

The solar thermal expanded through manufacture and deployment of solar water heating systems for individual homes, large systems for institutions, heaters and cookers all added up. The solar PV penetrated to the rural areas through the DC systems with 2 to 4 lights, fan etc catering to the basic needs of the backward areas. Many banks also supported these initiatives. This was also a period when a lot of consolidation happened in the industry.

From 2005 industry saw aggressive growth in the number of companies, business volume, products and spread. The national solar mission announcement excited the proponents of renewable industry and a large number of companies jumped the bandwagon. Many large solar PV projects are now rolled out throughout the country. Karnataka had a few pilot projects implemented with now Gujarat leading the pack followed by Rajasthan where solar irradiance is the highest,

Currently the industry is having about 150 in solar thermal and half of it in solar PV serious players of a total of 300 odd solar product providers through out the county. Business in India size wise is skewed towards the lower part and a rough estimate shows

Percentage of companies	Annual turnover (Crores of rupees)
2%	<50
10%	25 to 50
25%	15 to 25
About 65%	< 15

— Typical pattern of a nascent business which have a large localized content. The states leading are Karnataka, Maharashtra, Gujarat, New Delhi, Andrapradesh etc

Solar Products in India

1) Solar home systems DC

Technology	Solar Photovoltaic
Primary Usage / Target customers	Used in un electrified areas and area with highly unreliable power. Very high volume sales. Doing well in rural market. Supported by various banks including Grameen banks.
Deployed regions	Large parts of UP, Bihar, Maharashtra, Karnataka
Module/Collector capacity range	20 Wp to 60 Wp—one to 4 lights/ dc fan etc as load
Price band	Rs 10,000/—to Rs 40,000/-

2) Solar home systems AC

Technology	Solar Photovoltaic
Primary Usage / Target customers	Currently a niche and very low volume business. Upper middle class in cities & towns with unreliable power
Deployed regions	Karnataka, Maharashtra, West Bengal, New Delhi etc
Module/Collector capacity range	80Wp to 1000Wp—Can power any appliance at home
Price band	Rs 50,000/—to Rs 400,000/-

3) Solar Street light

Technology	Solar Photovoltaic
Primary Usage / Target customers	For pathways in parks, residential areas, individual homes etc
Deployed regions	Throughout India
Module/Collector capacity range	36 Wp to 100 Wp—These days good LED versions are available in the market
Price band	Rs 20,000/—to Rs 80,000/-

4) Solar Lanterns

Technology	Solar Photovoltaic
Primary Usage / Target customers	Works well as a replacement for Kerosene lamps. Un electrified deep rural market. A few in urban areas.
Deployed regions	Predominantly seen in Northern & Eastern parts of India
Module/Collector capacity range	2 Wp to 10 Wp
Price band	Rs 2000/—to Rs 4000/-

5) Solar water heaters—FPC

Technology	Solar Thermal
Primary Usage / Target customers	Individual homes using warm water for bathing
Deployed regions	Throughout India
Module/Collector capacity range	2 m² per collector—modular
Price band	Rs 20,000/—for a 100 Liters per day non pressurized system

6) Solar water heaters—ETC

Technology	Solar Thermal
Primary Usage / Target customers	Individual homes using warm water for bathing
Deployed regions	More seen in Southern & Western India
Module/Collector capacity range	1.7m² onwards
Price band	Rs 17,000/—for a 100 Liters per day non pressurized system

7) Commercial solar water heaters

Technology	Solar Thermal
Primary Usage / Target customers	Large Hotels, Hospitals, Hostels etc
Deployed regions	Throughout India
Module/Collector capacity range	1000 Liters Per day and above
Price band	Rs 110/—per litre for non pressurised system

8) Solar cookers

Technology	Solar Thermal
Primary Usage / Target customers	Community cooking
Deployed regions	Western & Northern India
Module/Collector capacity range	1.5 to 3 mtr dia dish
Price band	Rs 25,000/—to Rs 45,000/-

9) Solar water pumping

Technology	Solar PV
Primary Usage / Target customers	Remote water pumping for agriculture
Deployed regions	Scattered all over
Module/Collector capacity range	100 Wp to 500 Wp
Price band	3 lakhs to 8 lakhs

Initiatives in India to promote renewable energy:

Ministry of New and Renewable Energy (MNRE):

MNRE is the nodal ministry of the government of India. The aim of the ministry is to develop and deploy new and renewable energy for supplementing the energy requirements of the country. Very few countries around the world have such an exclusive organizational set up to support renewable energy. The administrative set up founded in 1981 has gone through transformations before becoming the present MNRE

- 1981—Commission for Additional Sources of Energy (CASE)
- 1982—Department of Non—conventional Energy Sources (DNES)
- 1992—Ministry of Non-conventional Energy Sources (MNES)
- 2006—Ministry of New & Renewable Energy (MNRE)

The mission of the ministry is to ensure:

- Increase in the share of clean power: Renewable (bio, wind, hydro, solar, geothermal & tidal) electricity to supplement fossil fuel based electricity generation
- Energy Availability and Access: Supplement energy needs of cooking, heating, motive power and captive generation in rural, urban, industrial and commercial sectors
- Energy Affordability: Cost-competitive, convenient, safe, and reliable new and renewable energy supply options
- Energy Equity: Per-capita energy consumption at par with the global average level by 2050, through a sustainable and diverse fuel—mix

MNRE operates many of its programmes through the various state nodal agencies. The MNRE head office is situated at New Delhi. MNRE along with the state nodal agencies will coordinate with the stakeholders including members of the industry, lending institutions, academicians, general public, and customers to bring out various programmes / schemes to promote the use of renewable energy.

State Nodal Agencies

Almost all states have state nodal agencies to support various programmes for promoting renewable energy.

The activities of SNA's include

- Implement state level programmes for renewable energy
- Advice state govt on renewable energy policies
- Act as implementing agency for MNRE programmes

- Interact with industry bodies
- Undertake demonstration projects and communicate educational and rural services
- Seek recommendation from technical experts on various subjects
- Review results and get feedback on projects implemented
- Ensure orderly release of funds and subsidies
- Promote research programme by sponsoring and coordinating
- Assess environmental effects of various programmes

State nodal agencies have a critical role to play in promotion of renewable energy to the grass roots. An efficient SNA can be very effective in creating a conducive environment for progression of renewable energy industry

Andhra Pradesh:
Non-Conventional Energy Development Corporation of Andhra Pradesh (NEDCAP) Ltd.
www.nedcap.gov.in

Arunachal Pradesh:
Arunachal Pradesh Energy Development Agency
www.apedagency.com

Assam:
Assam Energy Development Agency
www.assamrenewable.org

Bihar:
Bihar Renewable Energy Development Agency
www.sdabeebreda.com

Chhatisgrah:
Chhattisgarh State Renewable Energy Development Agency (CREDA)
www.credacg.com

Gujarat:
Gujarat Energy Development Agency (GEDA)
www.geda.org.in

Haryana:
Haryana Renewal Energy Development Agency (HAREDA)
www.hareda.gov.in

Himachal Pradesh:
HIMURJA,
http:/himurja.nic.in

Jharkhand:
Jharkhand Renewable Energy Development Agency
www.jreda.com

Karnataka:
Karnataka Renewable Energy Development Agency Ltd.
www.kredlinfo.in

Kerala:
Agency for Non-Conventional Energy and Rural Technology (ANERT),
http://anert.gov.in/

Madhya Pradesh:
MP Urja Vikas Nigam Ltd.,
www.mprenewable.org

Maharashtra:
Maharashtra Energy Development Agency (MEDA)
www.mahaurja.com

Manipur:
Manipur Renewable Energy Development Agency (MANIREDA)

Meghalaya:
Meghalaya Non-conventional & Rural Energy Development Agency
http://mnreda.gov.in

Mizoram:
Zoram Energy Development Agency (ZEDA)

Nagaland:
Nagaland Renewable Energy Development Agency

Orissa:
Orissa Renewable Energy Development Agency
www.oredaorissa.com

Punjab:
Punjab Energy Development Agency
http://peda.gov.in/eng/index.html

Rajasthan:
Rajasthan Renewable Energy Corporation Limited
http://www.rajenergy.com

Sikkim:
Sikkim Renewable Energy Development Agency,

Tamil Nadu:
Tamil Nadu Energy Development Agency (TEDA)
www.teda.gov.in

Tripura:
Tripura Renewable Energy Development Agency
http://treda.nic.in

Uttar Pradesh:
Non-conventional Energy Development Agency (NEDA), U.P.
http://neda.up.nic.in

Uttarkhand:
Uttrakhand Renewable Energy (UREDA) Development Agency

West Bengal:
West Bengal Renewable Energy Development Agency
www.wbreda.org

Reference : MNRE website

The list may not be complete. Many are added and updated.

AKSHAY URJA SHOPS

The Ministry has been promoting the establishment of Aditya Solar Shops in major cities of the country with a view to make solar energy products easily available and to provide after sales repair and maintenance services. The shops have been renamed as "Akshay Urja Shops" with a view to cover wider sales and service of all renewable energy devices and systems. It has been planned to establish one shop in each district of the country. A total of 302 shops in 31 States / UTs, (including 104 Aditya Solar Shops) have been established under the scheme. Under the Akshya Urja Shop Scheme, the network of the shops is being expanded by encouraging private entrepreneurs and NGOs to set up and operate such shops in all districts of the country. Applicants are eligible to avail loans up to Rs. 10 lakhs through designated banks for establishment of the shops at an interest rate of 7%. In addition, recurring grant & incentive linked with turnover upto Rs. 10,000 per month (subject to certain conditions) during the first two years of operation will also be available. The scheme will be operated through State Nodal Agencies and IREDA.

The Government of India has enacted several policies to support the expansion of renewable energy. A few of those other than JNSM are:

- Electricity Act 2003: Mandates that each State Electricity Regulatory Commission (SERC) establish minimum renewable power purchases; allows for the Central Electricity Regulatory Commission (CERC) to set a preferential tariff for electricity generated from renewable energy technologies; provides open access of the transmission and distribution system to licensed renewable power generators.
- National Electricity Policy 2005: Allows SERCs to establish preferential tariffs for electricity generated from renewable sources.
- National Tariff Policy 2006: Mandates that each SERC specify a renewable purchase obligation (RPO) with distribution companies in a time-bound manner with purchases to be made through a competitive bidding process.
- Rajiv Gandhi Grameen Vidyutikaran Yojana (RGGVY) 2005: Supports extension of electricity to all rural and below poverty

line households through a 90% subsidy of capital equipment costs for renewable and non-renewable energy systems.
- Eleventh Plan 2007-2012: Establishes a target that 10% of power generating capacity shall be from renewable sources by 2012 (a goal that has already been reached); supports phasing out of investment-related subsidies in favor of performance-measured incentives.

Roles played by banks in financing solar

The importance of the increasing need to provide renewable energy sources was recognised by the Indian government in the early 1970s. For nearly 10 years, India's significant efforts have gone into the design, development, field demonstration and large-scale use of a number of renewable energy products and systems.

After the Indian government in 1981 established a Commission for Additional Sources of Energy (CASE) to promote renewable energy and then on to MNRE in 2006 the focus of government has come a long way in supporting renewable energy sources deployment. India is probably the only country in the world that has an exclusive ministry which deals with new and renewable energy sources.

The 9th, 10th and 11th Five-Year Plans clearly indicated that rural development and energy are major goals for the plan periods. NABARD have actively supported these. But lot more can be done.

The largest barrier to the switch to solar in developing countries has been the lack of financing for clean energy in poor communities. A solar photovoltaics (PV) $1.5 million pilot project led by United Nations Environment Programme (UNEP), during the previous decade in India has transformed the lives of approximately 100,000 people living in poverty-stricken rural regions by providing several hours of uninterrupted lighting every night. During this project the communities had easier access to financing which provides an opportunity to pay more money upfront to acquire better, cleaner technologies that can save money in the long-term while improving the quality of life. Two of India's leading banks, Canara Bank and Syndicate Bank, were the project's original partners, jointly supplying low-interest loans that could be repaid over five years through their 2,000 rural branches.

With the project, the number of financed solar home systems in the pilot region of Karnataka state in south India increased from 1,400 in 2003 to over 18,000. The systems supply a few hours of continual power in homes or small shops to run small appliances and provide improved reading light.

Rural banks are proving to be instrumental in spreading sustainable energy. In Uttar Pradesh, one such example is that of Aryavart Gramin Bank (AGB), a regional rural bank, sponsored by the Bank of India. Many branches of the bank situated in rural areas experienced regular power disruption. This is one reason the prompted the bank to offer aid for solar systems came from their own brush with power shortage. Realising the success of their model, AGB targeted an ambitious 14.86 million villagers residing in 8,542 villages across 6 districts of UP. Through a financing scheme—'Ghar Ghar Me Ujala Saur Urja Se'—the bank offered to finance up to 85% of the project cost and opened its doors to households in rural, semi-urban and urban areas. Though the benefits of the scheme have largely been reaped by the rural population, it is not limited to village dwellers alone.

Karnataka Vikas Grameen converted 80 Indian villages into "solar villages", covering an estimated 10,000 houses under the project by the end of 2009. Under the project, KVGB is to cover 10,000 houses in these 80 villages by end-2009. The bank is provided finance for 85% of the cost of solar light units and buyers may choose the company for purchasing solar lights according to their choice.

The solar water heater systems were supported by banks through the subsidized loan schemes. Syndicate bank, Canara bank, Vijaya Bank etc have all during the last few years supported this scheme. The Bank of Maharashtra, a public sector company and the leading bank in the state of Maharashtra, has received several awards for their impressive commitment when it comes to disbursing loans and, thus, financing residential solar thermal systems in India. In the last nine years, the bank issued interest-reduced loans to 17,472 families to finance their solar water heaters.

Similarly the grameen banks are also active in supporting the deployment of solar systems in rural areas. However the solar industry in India need lot more support from the banking industry to establish a viable business.

Indian Renewable Energy Development Agency Ltd

- Indian Renewable Energy Development Agency Limited (IREDA) was established on 11th March, 1987 as a Public limited Government Company under the Companies Act, 1956

and it promotes, develops and extends financial assistance for Renewable Energy and Energy Efficiency/Conservation Projects.
- IREDA has been notified as a "Public Financial Institution" under section 4 'A' of the Companies Act, 1956 and registered as Non-Banking Financial Company (NFBC) with Reserve Bank of India (RBI).
- IREDA's mission is "Be a pioneering, participant friendly and competitive institution for financing and promoting self-sustaining investment in energy generation from Renewable Sources, Energy Efficiency and Environmental Technologies for sustainable development."
- IREDA's Motto is "Energy for Ever."

The main objectives of IREDA are:

1. To give financial support to specific projects and schemes for generating electricity and / or energy through new and renewable sources and conserving energy through energy efficiency.
2. To maintain its position as a leading organization to provide efficient and effective financing in renewable energy and energy efficiency / conservation projects.
3. To increase IREDA`s share in the renewable energy sector by way of innovative financing.
4. Improvement in the efficiency of services provided to customers through continual improvement of systems, processes and resources.
5. To strive to be competitive institution through customer satisfaction.
www.ireda.gov.in

SOLAR ENERGY CENTRE

The Solar Energy Centre (SEC), established in 1982, is a dedicated unit of the Ministry of New and Renewable Energy, Government of India for development of solar energy technologies and to promote its applications through product development

Activities:

- Solar Resource Assessment
- Solar Thermal

- Solar Buildings
- Solar Photovoltaics
- Solar Energy Materials
- Solar Thermal Power Generation
- Interactive Research & Development
- Bio-Fuel
- Consultancy
- International Co-operation
- Training & Information Services
- Visitors Programme

The Energy and Resources Institute (TERI):

TERI was formally established in 1974 with the purpose of tackling and dealing with the immense and acute problems that mankind is likely to face within in the years ahead on account of the gradual depletion of the earth's finite energy resources which are largely non-renewable and on account of the existing methods of their use which are polluting

Over the years the Institute has developed a wider interpretation of this core purpose and its application. Consequently, TERI has created an environment that is enabling, dynamic and inspiring for the development of solutions to global problems in the fields of energy, environment and current patterns of development, which are largely unsustainable. The Institute has grown substantially over the years, particularly, since it launched its own research activities and established a base in New Delhi, its registered headquarters. The central element of TERI's philosophy has been its reliance on entrepreneurial skills to create benefits for society through the development and dissemination of intellectual property. The strength of the Institute lies in not only identifying and articulating intellectual challenges straddling a number of disciplines of knowledge but also in mounting research, training and demonstration projects leading to development of specific problem-based advanced technologies that help carry benefits to society at large.

TERI has grown to establish a presence not only in different corners and regions of India but is perhaps the only developing country institution to have established a presence in North America and Europe and on the Asian continent in Japan, Malaysia and the Gulf.

The Institute established the TERI University in 1998. Initially set-up as the TERI School of Advanced Studies, it received the status of a deemed university in 1999. The University is a unique institution of higher learning exclusively for programmes leading to PhD and Masters level degrees. Its uniqueness lies in the wealth of research carried out within TERI as well as by its faculty and students making it a genuinely research based University.

www.teriin.org

Sardar Patel Renewable Energy Research Institute (SPRERI) was establishment way back in January 1979 as a result of the initiative taken by a group of foresighted persons led by Dr. H.M. Patel and Shri Nanubhai Amin.

Mission:

Develop technologies and provide services to promote the use and application of renewable energy.

SPRERI is a non-profit autonomous organization registered as a Society under the Societies Registration Act 21 of 1860 and also as a Public Trust under the Bombay Public Trust Act 1950.It has been approved as a Research Association for the purpose of clause (ii) of subsection (1) of Section 35 of the Income Tax Act 1961 and is recognized as a Scientific and Industrial Research Organization (SIRO) by the Department of Science & Technology, Govt. of India. It is recognized by S.P. University, Vallabh Vidyanagar as a Centre for Ph.D. research.

GOALS & OBJECTIVES

- Research and development of renewable energy systems and practices which are technically and economically sound and environment friendly and which can reduce the dependence of industry and rural sector on fossil fuels.
- Demonstration of renewable energy technology successfully integrated into the user systems
- Development and demonstration of techniques to reduce energy losses and energy intensity in processes and plants
- Acting as an independent agency for high quality testing of energy and environment protection systems.
- Acting as a store house of knowledge and published information and source of technical expertise/consultancy in the field of

renewable energy for the benefit of technology users, suppliers and policy makers
- Function as a nationally and internationally recognized institution for imparting training and education in renewable energy science and technology.
- Bring out technical publications and disseminate information on renewable energy technology through media and through popular literature, exhibitions, competitions and quizzes for students
- Periodically upgrade its infrastructure and other facilities and update its programmes to keep pace with the changes in renewable energy technology and economic and social environment.
- Develop and maintain effective arrangements to transfer RE technology for commercialization.
- Implementation of programmes and activities which support the above stated objectives.
- Develop cooperative/joint R&D programmes with reputed R&D organizations in the Country and outside.
www.spreri.org

ENVIS:

Introduction

Realising the importance of Environmental Information, the Government of India, in December, 1982, established an Environmental Information System (ENVIS) as a plan programme. The focus of ENVIS since inception has been on providing environmental information to decision makers, policy planners, scientists and engineers, research workers, etc. all over the country. ENVIS is a decentralised system with a network of distributed subject oriented Centers ensuring integration of national efforts in environmental information collection, collation, storage, retrieval and dissemination to all concerned.

Since environment is a broad-ranging, multi-disciplinary subject, a comprehensive information system on environment would necessarily involve effective participation of concerned institutions/ organisations in the country that are actively engaged in work relating to different subject areas of environment. ENVIS has, therefore, developed itself with a network of such participating institutions/organisations for the programme to be meaningful. A large number of nodes, known as ENVIS Centers, have been established in the network to cover the broad subject areas of environment with a Focal Point in the Ministry of Environment & Forests. These Centers have been set up in the areas of pollution control, toxic chemicals, central and offshore ecology, environmentally sound and appropriate technology, bio-degradation of wastes and environment management, etc.

Both the Focal Point as well as the ENVIS Centers has been assigned various responsibilities to achieve the Long-term & Short-term objectives. For this purpose, various services have been introduced by the Focal Point.

ENVIS due to its comprehensive network has been designed as the National Focal Point (NFP) for INFOTERRA, a global environmental information network of the United Nations Environment Programme (UNEP). In order to strengthen the information activities of the NFP, ENVIS was designated as the Regional Service Centre (RSC) of INFOTERRA of UNEP in 1985 for the South Asia Sub-Region countries . . . **ENVIS** focal point ensures integration of national efforts in environmental information collection, collation, storage, retrieval and dissemination to all concerned.

http://www.envis.nic.in/

SOLAR ENERGY SOCIETY OF INDIA (SESI):

The Solar Energy of India (SESI), established in 1976, and having its Secretariat in New Delhi, is the Indian Section of the International Solar Energy Society (ISES). Its interests cover all aspects of renewable energy, including characteristics, effects and methods of use, and it provides a common ground to all those concerned with the nature and utilization of this renewable non-polluting resource.

The Society is interdisciplinary in nature, with most of the leading energy researchers and manufacturers of renewable energy systems and devices of the country as its members. High academic attainments are not a prerequisite for membership and any person engaged in research, development or utilization of renewable energy or in fields related to renewable energy and interested in the promotion of renewable energy utilization can become a member of the society.

Organization

The Society is administered by its Governing a Council of twenty members elected once in two years, consisting of the President, six Vice Presidents, a Treasurer, a Secretary General and 11 members of whom one is the immediate past President. The council meets two or three times in year. The Annual General Meeting of the members is normally held at the time of the International Congress on Renewable Energy (ICORE).

Day to day administration is provided by the Society Secretariat headed by the Director General.

Regions in which sufficient interest exists, regional or local chapters of the Society have been/can be established.

Activities

The major activities of the Society are:

- Publication of SESI Journal, a bi-annual technical journal containing papers on renewable energy utilization, technical notes and other items of interest of those involved in renewable energy research and development.
- Publication of a monthly news letter, namely the SESI News Letter.
- Organization of one-day workshops on selected topics.
- Organization of the International Congress on Renewable Energy once in a year, where numerous scientific and technical papers are presented and discussed.
- Publication of the proceedings of the Annual Convention.

Website ; www.sesi.in

SEMI India

SEMI is the global industry association serving the manufacturing supply chains for the microelectronic, display and photovoltaic industries. SEMI member companies are the engine of the future, enabling smarter, faster and more economical products that improve our lives. Since 1970, SEMI has been committed to helping members grow more profitably, create new markets and meet common industry challenges.

SEMI is engaged in nearly all of the major technology regions of the world and maintains offices in Austin, Bangalore, Beijing, Brussels, Hsinchu, Moscow, San Jose, Seoul, Shanghai, Singapore, Tokyo, and Washington, D.C. Primary activities include conferences and trade shows, international standards development, public policy, market research, workforce development, environmental, health and safety (EHS) other industry advocacy.

SEMI India will oversee the region's development—and support member interest in India's burgeoning microelectronic and PV manufacturing supply chains. India introduced a new incentive scheme for solar power plants in January 2008 that could enable rapid market growth in the coming years. With state utilities mandated to buy energy from solar power plants, several state electricity regulatory boards are setting up preferential tariff structures.

Website—www.semi.org

Solar Thermal Federation of India (STFI)

STFI is a national association of solar water heater manufacturers representing more than 20 companies that together covers over 85% of today's solar water heater market. STFI serves as a voice of the solar water heater industry before government. The association aims to strengthen the performance of member companies, support industry growth and to work for the implementation of all steps to realize the high potential of solar water heater.

Objectives include:

- Improving interface with the government agencies
- Creating product standards
- Collecting and maintaining industry data
- Communicating and networking
- Human resources development
- Market development

Website—www.stfi.org.in

TERI BCSD:

Business Sustainability is the opportunity for business to improve its profitability, competitiveness, and market share without compromising resources for future generations. TERI-BCSD India (formerly called CoRE BCSD India) is a guide to the Indian corporate diaspora encouraging business people to develop a vision of a sustainable company, translate that vision into a management action plan and turn sustainability into a competitive advantage.

TERI-BCSD India is an independent and credible platform for corporate leaders to address issues related to sustainable development and promote leadership in environmental management, social responsibility, and economic performance (the triple bottom line). TERI-BCSD India is a partner of the WBCSD (World Business Council for Sustainable Development), Geneva and a member of its regional network. Presently, the network has a total of 84 corporate members across India representing a varied section of Indian industry. Subject experts from these member corporates identify and conceptualize projects and a team of industry members and TERI researchers then work to develop appropriate solutions and strategies for use by the industry. Workshops, trainings, seminars, events and publications are the other outreach activities of the business network.

Mission Statement

To provide an independent and credible platform for corporate leaders to address issues related to sustainable development and promote leadership in environmental management, social responsibility, and economic performance.

Website:bcsd.teri.res.in

Source: Respective Websites

Solar cities:

Urbanization and economic development are leading to a rapid rise in energy demand in urban areas. Urban areas have emerged as one of the biggest sources of Green House Gas (GHG) emissions, with buildings alone contributing to around 40% of the total GHG emissions. Urban areas are heavily dependant on fossil fuels, for the maintenance of essential public services, for powering homes, transport systems, infrastructure, industry and commerce.

As per latest UN report one million people are moving to urban areas each week. It is estimated that around two-thirds of the world population will be living in cities in 2050. This requires a tremendous shift in energy resources in urban areas.

Several Indian cities and towns are experiencing 15% growth in the peak electricity demand.

The Renewable Energy Ministry has already initiated various programmes in the Urban Sector for promoting solar water heating systems in homes, hotels, hostels, hospitals and industry; deployment of SPV systems/devices in urban areas for demonstration and awareness creation. The Energy Resources Institute (TERI), Various Industry Bodies, Associations are raising awareness and promotion of energy efficient Solar/Green Buildings. Bureau of Energy Efficiency under Ministry of Power have launched Energy Conservation Building Code (ECBC) which is aimed at energy efficiency measures and installation of renewable energy systems/devices in buildings including solar water heating systems. The programme on "Development of Solar Cities" would support/encourage Urban Local Bodies to prepare a Road Map to guide their cities in becoming 'renewable energy cities' or 'solar cities' or 'eco/green cities'.

What is a Solar City?

The Solar City aims at minimum 10% reduction in projected

demand of conventional energy at the end of five years, through a combination of enhancing supply from renewable energy sources in the city and energy efficiency measures. The basic aim is to motivate the local Governments for adopting renewable energy technologies and energy efficiency measures. In a Solar City

All types of renewable energy based projects like solar, wind, biomass, small hydro, waste to energy etc. may be installed along with possible energy efficiency measures depending on the need and resource availability in the city

Objectives:

- To enable and empower Urban Local Governments to address energy challenges at City—level.
- To provide a framework and support to prepare a Master Plan including assessment of current energy situation, future demand and action plans.
- To build capacity in the Urban Local Bodies and create awareness among all sections of civil society.
- To involve various stakeholders in the planning process.
- To oversee the implementation of sustainable energy options through public—private partnerships.

A total of 60 cities/towns are proposed to be developed as "Solar Cities" during the 11th Plan period. At least one city in each State to a maximum of five cities in a State will be supported by the Ministry. The cities included in the program will have more than 0.5 Million and less than 5 Million population. Relaxation could be considered in the case of special category state including North-Eastern States.

List of 48 Solar cities to be developed in India

State	Cities for which in-principle approval given
Andhra Pradesh	1. Vijayawada
Assam	2. Guwahati
	3. Jorhat
Arunachal Pradesh	4. Itanagar
Chandigarh	5. Chandigarh
Chhattisgarh	6. Bilaspur
	7. Raipur
Gujarat	8. Rajkot
	9. Gandhinagar
	10. Surat
Goa	11. Panji City
Haryana	12. Gurgaon
	13. Faridabad
Himachal Pradesh	14. Shimla
	15. Hamirpur
Karnataka	16. Mysore
	17. Hubli-Dharwad
Kerala	18. Thiruvananthapuram
	19. Kochi

State	City
Maharashtra	20. Nagpur
	21. Thane
	22. Kalyan-Dombiwali
	23. Aurangabad
	24. Nanded
	25. Shirdi
Madhya Pradesh	26. Indore
	27. Gwalior
	28. Bhopal
	29. Rewa
Manipur	30. Imphal
Mizoram	31. Aizawl
Nagaland	32. Kohima
	33. Dimapur
Orissa	34. Bhubaneswar
Punjab	35. Amritsar
	36. Ludhiana
	37. SAS Nagar (Mohali)
Rajasthan	38. Ajmer
	39. Jaipur
	40. Jodhpur
Tamil Nadu	41. Coimbatore
Tripura	42. Agartala
Uttrakhand	43. Dehradun
	44. Haridwar&Rishikesh
	45. Chamoli-Gopeshwar
Uttar Pradesh	46. Agra
	47. Moradabad
West Bengal	48. Howrah

A SWOT analysis of the Solar power in India:

Strengths	Weakness
<ul><li>Solar irradiance</li><li>An exclusive central ministry</li><li>An experienced and vibrant Industry and associated bodies</li></ul>	<ul><li>Low Awareness</li><li>High Cost</li><li>Low R&D efforts</li><li>Few world class manufacturing setup</li><li>Not a mainstream business yet</li></ul>
Opportunities	**Threats**
<ul><li>Large Population</li><li>High growth economy</li><li>Scarcity of electric power</li><li>Job creation</li></ul>	<ul><li>Policy implementation hurdles</li><li>High cost in product certification</li><li>Unhealthy competition</li></ul>

Looking Forward:

India has tripled its renewable energy generation capacity in the last five years and now rank fifth in the world in total installed renewable energy capacity and has established a legal and regulatory frame work to oversee this industry. The government plans to increase the capacity to generate renewable energy to 55 GW by the end of 13th Five year plan (2022) and the national action plan on climate change proposes to have renewable energy generation increase by 1% annually. The Jawaharlal National Solar Mission (JNSM) seeks to increase the combined solar capacity from to 20GW by 2022.

If India continues to grow at the rate of 8% for the next 10 years, the country's demand for power is set to soar from around 200 GW at present to 315 to 335 GW by 2017. Four factors will drive this demand.

- India's manufacturing growth
- Residential consumption growing at 14%
- The connection of 125,000 villages to the grid
- The realisation of the demand suppressed by load shedding

To support the above growth with reduced emission of pollutants, it is imperative that solar energy utilization should grow to a very high level. The On Grid PV market and deployment will now grow at the highest pace supported by not only JNSM but also enthusiastic state governments. The off grid is lying low and will continue its slow growth in the next couple of

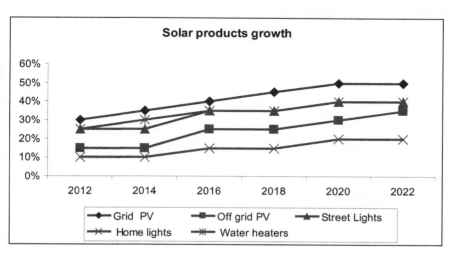

Years but with signs of good potential. By 2015 it will surpass the 20% growth and accelerate faster. The solar water heater which is a established market continue to exercise its dominance contributing 30 to 35% growth as more and more states start adopting it. Street lights category is one of the sleeping giants and is now gaining acceptance with increase in deployment. This will continue to grow at over 30% by 2018. The DC home lights will continue to grow at a slower pace in regional pockets. In terms of qty another segment that can multiply aggressively is the mini PV—sub Rs 5000/—, The Lanterns, Plug and Plays, The Micro Back up etc. This market will depend on the innovative products coming out and its reach in the established distribution chain.

Currently the solar deployment is anchored around the govt projects now baring residential solar water heaters. The Govt will continue to the largest buyers of these products. Now many products will generate enthusiasm in private players and this will stimulate the market. The 10Kwp Solar PV, Streetlights, Solar wind Hybrid, Process heating all have good potential. The banking industry which is now staying away need to come in with simplified regulations.

The quality of deployment and product up gradation / differentiation should become the key strategies for the industry rather than the mad rush of somehow offering the lowest price which is both detrimental to the industry and end customer.

Considering the immediate target for solar is 20 GW and current installed capacity has only reached one GW with the entire solar heat excluded—it is a long way up.

CHAPTER 9

NATIONAL SOLAR MISSION

> *"Our vision is to make India's economic development energy-efficient. Over a period of time, we must pioneer a graduated shift from economic activity based on fossil fuels to one based on non-fossil fuels and from reliance on non-renewable and depleting sources of energy to renewable sources of energy. It would also enable India to help change the destinies of people around the world."*
>
> *Dr. Manmohan Singh*
> *Prime Minister of India*

Overview:

National solar mission named after our 1st prime minister Pandit Jawaharlal Nehru is another milestone in India's journey towards building a sustainable society to ensure the growth in economy with minimum or no impact on climate change and is an outcome of the programme and policies followed by our successive governments and now kick started in 2007.

5th June 2007—The Prime Minister, Dr. Manmohan Singh, set's up a High Level advisory group on climate change issues named **"Prime Minister's council on climate change"**. The team includes senior Government officials from related ministries and well known persons from the scientific community.

13th July 2007—The council met and agreed to prepare a national document compiling the action taken by India for addressing the challenge of climate change and action it propose to take.

30th June 2008—National Action plan on climate change (NAPCC) prepared by the Prime Ministers council on climate change released by the Prime Minister outlining National Solar Mission as one of the eight national missions laid out in the document.

11th January 2010—National solar mission Launched by Prime Minster as prepared by Ministry of New and Renewable energy

National Action Plan on Climate Change:

This document is aimed to address the India's challenge of sustaining its rapid economic growth while dealing with global threat of climate change.

Vision—To create a prosperous, but not wasteful society, an economy that is self sustaining in terms of its ability to unleash the creative energies of our people and is mindful of our responsibilities to both present and future generations.

Objective—To establish and effective, cooperative and equitable global approach based on principles of common but differentiated responsibilities and respective capabilities, enshrined in the United Nations Framework Convention on Climate Change.

Approach & Way Forward—Eight National Missions form the core of National Action plan

- National Solar Mission
- National Mission for Enhanced Energy Efficiency
- National Mission on Sustainable Habitat
- National Water Mission
- National Mission for Sustaining the Himalayan Ecosystem
- National Mission for a Green India
- National Mission for Sustainable Agriculture
- National Mission on Strategic Knowledge for Climate Change

Implementation—The national missions will be institutionalized by the respective ministries organized through inter—sectoral groups which include in addition to related ministries, the finance ministry, planning commission, experts from industry, academia and civil society.

National Solar Mission:

The National Solar Mission is a major initiative of the Government of India and State Governments to promote ecologically sustainable growth while addressing India's energy security challenge. It will also constitute a major contribution by India to the global effort to meet the challenges of climate change.

The National Action Plan on Climate Change also points out: "India is a tropical country, where sunshine is available for longer hours per day and in great intensity. Solar energy, therefore, has great potential as future energy source. It also has the advantage of permitting the decentralized distribution of energy, thereby empowering people at the grassroots level".

Based on the above a National Solar Mission is being launched under the brand name "Solar India".

Responsibility:

a) The deployment of commercial and non commercial solar technologies in the country
b) Establishing a solar research facility at an existing establishment to coordinate the various research, development and demonstration activities being carried out in India, both in public and private sector
c) Realizing integrated private sector manufacturing for solar material, equipment, cells and modules
d) Networking of Indian research efforts with international initiatives
e) Providing funding support for the above activities

Objectives: The objective of the National Solar Mission is to establish India as a global leader in solar energy, by creating the policy conditions for its diffusion across the country as quickly as possible thereby developing a solar industry that is capable of delivering solar energy competitively against fossil options for small to large scale electric power requirement in the span of 20 to 25 years.

Mission Targets

The Mission will adopt a 3-phase approach, spanning the remaining period of the 11th Plan and first year of the 12th Plan (up to 2012-13) as Phase 1, the remaining 4 years of the 12th Plan (2013-17) as Phase 2 and the 13th Plan (2017-22) as Phase 3.

At the end of each plan, and mid-term during the 12th and 13th Plans, there will be an evaluation of progress, review of capacity and targets for subsequent phases, based on emerging cost and technology trends, both domestic and global.

The deployment target across the application segments is envisaged as follows:

	Phase-1 2010-2013	Phase-2 2013-2017	Phase-3 2017-22
Solar Collectors (Sq mtrs)	7 million	15 million	20 million
Off grid solar applications (MW)	200	1000	2000
Utility grid power, including roof top (MW)	1000-2000	4000-10000	20000
Solar lighting systems for rural areas			20 million

Mission strategy:

The first phase will announce the broad policy frame work to achieve the objectives of the National Solar Mission by 2022. The policy announcement will create the necessary environment to attract industry and project developers to invest in research, domestic manufacturing and development of solar power generation and thus create the critical mass for a domestic solar industry. The Mission will work closely with State Governments, Regulators, Power utilities and Local Self Government bodies to ensure that the activities and policy framework being laid out can be implemented effectively. Since some State Governments have already announced initiatives on solar, the Mission will draw up a suitable transition framework to enable an early and aggressive start-up.

Utility connected applications: constructing the solar grid

The key driver for promoting solar power would be through a Renewable Purchase Obligation (RPO) mandated for power utilities, with a specific solar component. This will drive utility scale power generation, whether solar PV or solar thermal. The Solar Purchase Obligation will be gradually increased while the tariff fixed for Solar power purchase will decline over time.

The below 80°C challenge—solar collectors

The Mission in its first two phases will promote solar heating systems, which are already using proven technology and are commercially viable. The Mission is setting an ambitious target for ensuring that applications, domestic and industrial, below 80 °C are done using solar. The key strategy of the Mission will be to make necessary policy changes to meet this objective:

· Firstly, make solar heaters mandatory, through building byelaws and incorporation in the National Building Code, ·

Secondly, ensure the introduction of effective mechanisms for certification and rating of manufacturers of solar thermal applications ·

Thirdly, facilitate measurement and promotion of these individual devices through local agencies and power utilities, and ·

Fourthly, support the upgrading of technologies and manufacturing capacities through soft loans, to achieve higher efficiencies and further cost reduction.

The off-grid opportunity—lighting homes of the power—deprived poor:

A key opportunity for solar power lies in decentralized and off-grid applications. In remote and far-flung areas where grid penetration is neither feasible nor cost effective, solar energy applications are cost-effective. They ensure that people with no access, currently, to light and power, move directly to solar, leap-frogging the fossil fuel trajectory of growth. The key problem is to find the optimum financial strategy to pay for the high-end initial costs in these applications through appropriate Government support.

Currently, market based and even micro-credit based schemes have achieved only limited penetration in this segment. The Government has promoted the use of decentralized applications through financial incentives and promotional schemes. While the Solar Mission has set a target of 1000 MW by 2017, which may appear small, but its reach will add up to bringing changes in millions of households. The strategy will be learn from and innovate on existing schemes to improve effectiveness.

Manufacturing capabilities: innovate, expand and disseminate

Currently, the bulk of India's Solar PV industry is dependent on imports of critical raw materials and components—including silicon wafers. Transforming India into a solar energy hub would include a leadership role in low-cost, high quality solar manufacturing, including balance of system components. Proactive implementation of Special Incentive Package (SIPs) policy, to promote PV manufacturing plants, including domestic manufacture of silicon material, would be necessary.

Indigenous manufacturing of low temperature solar collectors is already available; however, manufacturing capacities for advanced solar collectors for low temperature and concentrating solar collectors and their components for medium and high temperature applications need to be built. An incentive package, similar to SIPS, could be considered for setting up manufacturing plants for solar thermal systems/ devices and components.

The SME sector forms the backbone for manufacture of various components and systems for solar systems. It would be supported through soft loans for expansion of facilities, technology upgradation and working capital. IREDA would provide this support through refinance operations. It should be ensured that transfer of technology is built into Government and private procurement from foreign sources.

R&D for Solar India: creating conditions for research and application

A major R&D initiative to focus:

Firstly, on improvement of efficiencies in existing materials, devices and applications and on reducing costs of balance of systems, establishing new applications by addressing issues related to integration and optimization;

Secondly, on developing cost-effective storage technologies which would address both variability and storage constraints, and on targeting space intensity through the use of better concentrators, application of nano-technology and use of better and improved materials.

The Mission will be technology neutral, allowing technological innovation and market conditions to determine technology winners.

R&D strategy would comprise dealing with five categories viz.

i) Basic research having long term perspective for the development of innovative and new materials, processes and applications,
ii) Applied research aimed at improvement of the existing processes, materials and the technology for enhanced performance, durability and cost competitiveness of the systems/ devices,
iii) Technology validation and demonstration projects aimed at field evaluation of different configurations including hybrids with conventional power systems for obtaining feedback on the performance, operability and costs,
iv) Development of R&D infrastructure in PPP mode
v) Support for incubation and start ups.

A Solar Research Council will be set up to oversee the strategy, taking into account ongoing projects, availability of research capabilities and resources and possibilities of international collaboration.

A National Centre of Excellence (NCE) shall be established to implement the technology development plan formulated by the Research Council and serve as its Secretariat. It will coordinate the work of various R&D centers, validate research outcomes and serve as an apex centre for testing and certification and for developing standards and specifications for the Solar industry. It is envisaged that the Solar Energy Centre of the MNRE will become part of the National Centre of Excellence.

The Research Council, in coordination with the National Centre of Excellence, inventorize existing institutional capabilities for Solar R&D and encourage the setting up of a network of Centers of Excellence, each focusing on an R&D area of its proven competence and capability.

Human Resource Development

An ambitious human resource development programme, across the skill-chain, will be established to support an expanding and large-scale solar energy programme, both for applied and R&D sectors. In Phase I, at least 1000 young scientists and engineers would be incentivised to get trained on different solar energy technologies as a part of the Mission's long-term R&D and HRD plan.

The rapid and large-scale diffusion of Solar Energy will require an increase in technically qualified manpower of international standard. It is envisaged that at the end of Mission period, Solar industry will employ at least 100,000 trained and specialized personnel across the skill spectrum. These will include engineering management and R&D functions.

The following steps may be required for Human Resource Development:

- IITs and premier Engineering Colleges will be involved to design and develop specialized courses in Solar Energy, with financial assistance from Government. These courses will be at B. Tech, M. Tech and Ph. D level.
- A Government Fellowship programme to train 100 selected engineers / technologies and scientists in Solar Energy in world class institutions abroad will be taken up. This may need to be sustained at progressively declining levels for 10 years.
- Setting up of a National Centre for Photovoltaic Research and Education at IIT, Mumbai drawing upon its Department of Energy Science and Engineering and its Centre for Excellence in Nano-Electronics.

Implementation Arrangements:

The Scheme would be implemented through multiple channel partners for rapid up-scaling in an inclusive mode. It is envisaged that these channel partners would enable significant reduction in transaction cost and time, since without these arrangements, individuals and small groups of clients may not be in a position to access the provisions of the scheme. Channel partners which would be used for implementation could include the following:-

a) Renewable Energy Service Providing Companies (RESCOs)
b) Financial Institutions including microfinance institutions acting as Aggregators
c) Financial Integrators
d) System Integrators
e) Programme Administrators

a) Renewable Energy Service Providing Companies (RESCOs):

These are companies which would install, own & operate RE systems and provide energy services to consumers. These entities may tie up with FIs for accessing the financial support under the scheme.

b) FI s including MFIs acting as Aggregators:

These would be institutions which are involved in consumer finance and have established base of customers in rural/urban areas and outreach through self help groups, etc.

These would typically access interest subsidy through refinance facility as also credit linked capital subsidy on behalf of their borrowers from IREDA.

c) Financial Integrators:

These are entities which would integrate different sources of finance including carbon finance, government assistance and other sources of funds to design financial products/ instruments and make these available to their clients at an affordable cost. These entities would tie up with manufacturers and service providers.

d) System Integrators:

These are companies/ entities which would provide RE systems & services to clients including design, supply, integration and installation, O&M and other services. These entities may tie up with FIs for accessing the financial support under the scheme.

e) Programme Administrators:

These would include, inter alia, Central and State Government Ministries and Departments and their organizations, State Nodal Agencies, Utilities, Local bodies, PSUs and reputed Non-Governmental Organizations (NGOs). These entities would directly implement the scheme and access capital subsidy (non credit linked) from MNRE.

The various channel partners who can participate in this Scheme have been described above and a transparent methodology for accrediting these entities by MNRE is in place. The parameters for accrediting an entity could comprise of:

a) Net worth / turnover of the participating entity
b) Technical capability for carrying out services which would, inter alia, include site selection, feasibility study, design, value engineering, cost optimization, time scheduling, procurement, installation/commissioning and O&M functions
c) Credit rating, if any
d) Track record
e) Tie-ups with equipment providers.

The accreditation process would categorize the various entities into grades which would determine the quantum of work in terms of financial limits that they could undertake under the Scheme. This accreditation process would also enable inclusion of start ups with the requisite technical and installation skills. An opportunity would be provided for young entrepreneurs to participate as channel partners in order to tap their creative potential as innovators.

The Mission document and various addendums/ annexure released covers all the criteria and requirements including technical requirements, boundary conditions for various categories including Solar PV and Thermal applicable for different users, Minimum technical requirements, Standards, Specifications, Warranty of products supplied etc

More detailed and well laid out information is available in the MNRE website. The readers/ Customers interested in understanding more may refer to the MNRE website or request to the system integrator.

Subsidy for Solar Products (May change—refer latest):

Solar PV systems with Battery	Around Rs 100/—per Wp
Solar PV systems without Battery	Around Rs 70/—per Wp
Solar Water heaters with Evacuated Tube Collector	Rs 3000/—per Sq mtr of collector area
Solar Water heaters with Flat Plate Collector with Water as Working fluid	Rs 3300/—per Sq mtr of collector area

Solar Water heaters with Flat Plate Collector with Air as Working fluid	Rs 2400/—per Sq mtr of collector area
Solar collector system for direct heating applications	Rs 3600/—per Sq mtr of collector area

. The capital subsidy/ unit collector area, as given above, is based on 30% of the benchmark costs, which would be reviewed annually. Capital subsidy would be computed based on the applicable type of solar collector/ modules capacity multiplied by the collector area/ no of modules involved in a given solar thermal application/project and is applicable for residential and commercial customers.

Besides the capital subsidy as proposed above, the pattern of support could include a soft loan at 5%, as under:

Soft loan @ 5% interest would be available, inter alia, for balance cost which may comprise installation charges, cost of civil work for large systems and costs of accessories (viz. insulating pipeline, electric pump, controllers and valves, additional water tanks, blower for air heating systems, drying trays for solar dryers, steam system, etc.), etc.

This translates to the end customer receiving the solar product at approximately 70 % of the listed MRP plus an option of receiving a soft loan for the balance amount subject to guidelines set by the lending institution.

A few examples are

- Solar water heater costing Rs 18000/—earlier is now available at 13000/—
- Solar water heater costing Rs 24000/—earlier is now available at 17000/—
- Solar PV DC home system costing Rs 40000/—is now available at Rs 28,000/—
- Solar PV AC home system costing Rs 2lacs is now available at Rs 1.4 lacs
- Solar PV AC home system costing Rs 5lacs is now available at Rs 3.8 lacs

Similar is the case of Solar cookers, Solar water pumping etc. The exact amount will vary in accordance with the benchmark cost set by the MNRE from time to time and the price list of the manufacturer.

In addition to above the individual customers will benefit from financial support announced by various state authorities wherever applicable. All the information under subsidy are indicative only. The actual subsidy at the time of purchase may be understood clearly from your reliable nearest vendor or authorities concerned.

The way forward for retail SHS support:

Quality power/ Energy are now a basic human need. Its availability is a must for a dignified living. For our country that is recording good consistent growth year after year and marching towards developed world, energy security for all is a prime requisite for inclusive growth or else the result will be disparity all-round with pockets of prosperity and poverty. It is beyond doubt now that renewable energy, especially solar can augment the power availability in India. The central and state governing establishments should create and promote an environment that will develop and sustain a robust renewable energy industry which is accessible to common man. In this contest the JNSM in its present form is inadequate to propel the installation of solar home systems countrywide. Hence three scenarios that have good potential to support solar are looked here.

Scenario A:

Currently under the JNSM government has subsidized the solar power equipments. This has led to a large number of MW scale projects being rolled out for harnessing solar energy. This power is fed to the grid where the distribution and transmission losses are high sending a large portion of subsidies directly into the drain. But on the other hand feeding the grid with solar is the only way now to use the existing infrastructure for power distribution. However solar being available throughout the country and easy to harness should now be vigorously promoted to be installed for individual homes. Once successful, this will have a significant impact on the energy security of India. The current JNSM should carve out specific target in this segment and run separate programme to boost decentralized harnessing of solar energy. The disbursement of subsidy should be simple but targeted to reach the beneficiary on time without hurting the stakeholders. Significantly more awareness creation is a must to get the common man in.

Scenario B:

Adequately sized solar installation for individual home requires reasonably high initial investments. Even though the recurring expenses are minimal, this high initial cost is a deterrent for many home owners looking for solar. A good financing environment will definitely help to overcome this. It is proved beyond doubt that the real estate sector especially individual housing recorded phenomenal growth during the last two decades from the support of good financing mechanism. The banking industry evolved itself to support the real estate market and made good business. Had this not happened, the real estate sector would not have seen the progress it witnessed so far. Next it was the turn of automobiles especially passenger cars which saw good growth in the last decade. Here also availability of financing is a major plus.

Similarly for solar products, a robust financing environment is required to propel its growth. The central government through Reserve Bank of India should promulgate policies to drive loans for clean energy investment in individual homes. Currently solar financing is not a serious business in banking industry even though many banks vouch for it. This is a sad part of the thirty years of solar products in India. The prospective customers and business executives of solar companies struggle to get through loan for solar products. More so for PV. A few bank branches are exceptions. This trickle financing is severely effecting the growth of solar installations for individual homes. The solar financing portfolio of banks needs to become big. The government should support this business segment and may be incubate it through well laid out policy guidelines. The incentives should be good enough for banks to concentrate and grow this now nascent banking business.

Scenario C:

Distribution of solar systems and its countrywide acceptance will have significant impact on the society. Hence it is prudent for government to get involved in its countrywide distribution akin to the existing Public Distribution System (PDS) for food supplies. This means a newly constituted govt / quasi govt organization will procure solar system components like solar panel, inverter, battery, charge controller etc from manufacturers and ensure its availability throughout India from where the local integrators can source and sell to the customers. There should be laid down standards and specifications to ensure quality of products—assessed and overseen by competent authority within the organization. Even though this will currently look impossible, strong measures in this direction will ensure the creation of a transparent and efficient authority which will reinforce the consumer confidence in the products and industry stakeholders can focus on manufacturing, sales and service. The certification and logistics should be taken care freely by the government organization that has very good time bound work execution capacity. With high consumer confidence, certified products, no logistics hassle and an extremely enticing market, the solar companies should be able to do good business and accelerate the installation of solar home systems throughout the country. Companies with strength in manufacturing can sell their products to the Solar PDS, but no tendering or price control—good products get good price. Companies with strength in sales and service can get certified products from Solar PDS to the customers—products with government guarantee matching with the current grid power.

The footnote is, the concept and targets of JNSM deserves accolades and implementation deserves good support. However much more need to be done fast for an emerging nation reeling under acute power shortage and threat of increased pollution but every year receiving 5000 trillion units of free solar energy with over 250 to 300 sunny days.

CHAPTER 10

MOVE OFF THE GRID

> *"Will we look into the eyes of our children and confess,
> That we had the opportunity, but lacked the courage?, That
> We had the technology, but lacked the vision?"*
>
> —*Green peace*

A sustainable society can only be build through sustainable development. Sustainable society is the one that lasts for generations and with its wisdom prevent destruction or undermine the physical and social system it lies on.

Sustainable development is defined as an integral economic, technological, social and cultural development adjusted to the needs of the environment protection, which enables current and future generations to meet their needs and improve their life quality. Sustainable development is focused on the preservation of the natural eco-systems and on the rational use of the natural treasures of the Earth. In other words Sustainable development implies nature preservation by man on sustainable basis and its use to the extent of its reproduction. Overuse and uncontrolled use and exploitation of the natural resources can cause a violation of the ecological balance and thus ecological disasters as well.

The basic principle of sustainable development is that natural resources can be exploited only to the level that provides their reproduction.

The planet is on the brink of runaway climate change. If annual average temperatures rise by more than 2ºC, the entire world will face more natural disasters, hotter and longer droughts, failure of agricultural areas and massive loss of species. Since this climate change is caused by burning fossil fuels, we urgently need an energy revolution, changing the world's energy mix to a majority of non-polluting sources. To avoid dangerous climate change, global emissions must peak in 2015 and start declining thereafter, reaching as close to zero as possible by mid-century

The cleanest source of energy is the Sun that for 5 billion years has been enabling and sustaining life on Earth. Sun energy through the process of photosynthesis is used for the growth and development of plants. We have seen in the previous chapters that sun's both form of energy—Light and Heat—can be processed directly and indirectly to provide useful energy for human living. This energy received from the Sun can offset the use of fossil fuel energy and thereby promote sustainable development.

India is undoubtedly on the path of high GDP growth clocking consistently more than 7% for the last few years and promises to become a major player in the global economy. It is time that we the people living in this great country demonstrate our commitment for a sustainable society. Our home is the right place to start concrete action in this regard. While simplest way to start reduce the energy consumption is by improving energy efficiency, people who can afford should opt for solar power and move away from the state board electricity power. Just imagine how wonderful it is to generate at home the power required for your use.

I have heard many people make passing remark that solar is costly. May be yes—May be not. To get it on the right perspective let us see a few things that we buy these days.

A look at what people in India are now spending annually for from published figures (cost of products in INR)

1. No of cars > 4 lakhs sold = 2 million nos
2. No of motorcycles > 0.5 lakh sold = 10 million nos
3. No of cars > 10 lakhs sold = 0.3 million nos
4. No of Super bikes costing over 8 lakhs =0.05 million nos

And estimated figures suggest

5. No of new homes > 20 lakhs sold = 1 million nos
6. No of Kitchen beautification >2 lakhs sold = 2 million nos
7. No of people spending on air travel > 2lakhs = 0.1 mln nos

8. No of people spending on Jewellery purchase > 5 lakhs = 0.1 million nos
9. No of new gadget purchase > 1 lakhs = 0.4 million nos
10. No of people spending on leisure > 2 lakhs = 0.2 mln nos

And this high scale spending list goes on!

Now what is the story of Solar / Renewables?

No of people buying Solar home system > 2 lakhs < 0.001 mln nos
It is miniscule in all sense.

Definitely cost alone is not the reason for such a disparity on preferred purchase or consumer spending. It is the conviction of the people and the right environment that is missing here.

A silver line in the horizon is the recent buoyancy shown by the renewable energy industry in India and worldwide and a few initiatives by the union and state governments. This has resulted in solar products and services available at a more reasonable price and prompted many among us to at least make plans to install it in our homes.

What is the real cost / qty of power we consume?

Generation to home

The power generated at the power plant goes through different stages of fuel processing, storage, distribution etc before it reach our home. Every activity/ process here has its efficiency and related losses. Hence power from power plant to home will have these minimum losses. These losses need to be considered in assessing what is the actual primary energy consumed to generate 1 unit of energy for our use when fed through the grid.

Power plant efficiency—35%
Storage/No load losses—20%
Transmission & Distribution losses—35%
Energy efficiency of the load at our home—70%

Added to this are the upstream recurring inefficiencies for all fossil fuels in exploration, extraction, transportation, refining etc.

A simple calculation will reveal that for getting 1 unit of usable energy for our home, about 8 to 10 units of primary energy should be consumed. Here the economic, environmental and negative social impacts get multiplied several times even though our usage is assumed to be minimal. This is the real cost of power that we now use from the grid.

Renewables, especially solar should soon become one among the top 5 requirements of every Indian household—what ever be the size of the solar home system. And here India with its vast High income class & Middle class population have a golden chance to beat every other country in the world of having installed the highest no of individual solar home systems with our country lying in the equatorial belt justifies this. Buying / Installing a renewable energy system is a powerful and direct way for you to help protect the environment, and make a long-lasting commitment to our planet's future.

Why Solar is the best solution

The amount of energy the sun sends towards our planet is thousands of times more than what we currently produce and consume. Some part of this energy—better known as solar radiation—is reflected back into space but a lot of it is absorbed by the atmosphere and other elements surrounding the inner atmosphere. We have seen that this energy can be easily harnessed for practical purposes such as heating homes, lighting bulbs and running automobiles and even airplanes. The uses can be as varied as the uses of energy itself. And the great thing is that we are never going to run out of this massive energy resource even for thousands of years.

As understood earlier Solar energy can be generated in two forms, namely electricity and heat. Solar cells or "photovoltaics" are used to convert solar radiation into electricity. Photovoltaic systems release no greenhouse gases into the atmosphere and they don't even need direct sunlight to produce energy; they just need sufficient daylight and this means environment friendly power from sunlight.

Electricity is generated indirectly too by first generating heat from solar energy and then using the steam produced in the process to run power generators. Here too, since no fossil fuels are being burned to produce heat, the resultant energy to 100% eco-friendly.

The indispensable Solar energy—*Random Thoughts*

Solar energy is extremely important for us to survive. Besides helping us to stay warm, and enabling other organisms to survive, it is used in the commercial power production as well.

Solar energy is also essential for plant life as it is necessary for photosynthesis, the process through which plants generate energy and process nutrients for their growth by converting solar light for their own use by utilizing chloroplasts within their leaves and bodies. Through this process, food is produced for other herbivores together with oxygen for us to breath and removal of carbon dioxide from the atmosphere, allowing creatures such as ourselves to survive while maintaining a balanced world temperature.

Solar energy is used for agriculture. Solar powered greenhouse farming is becoming more and more popular. Power from the sun will help the plants to grow and thrive. Greenhouses also take in the rays of the sun to heat up the inside of the greenhouse. This is very beneficial to the plants during the winter time.

Solar energy is also responsible for generating wind currents, which help the spreading of spores and gametes in plants and fungi to allow their propagation over a wide area.

Herbivores or omnivores also require solar energy to gain energy, which help them to produce various B vitamins such as B-12 to energize and stimulate several processes.

Solar energy even affects our daily moods, as our bodies adapt and adjust throughout the different seasons as the amount of sunlight changes. For example, many people feel more depressed or less energetic in the winter or during rainy seasons, while the summer tends to see more positive feelings and productive work styles.

Solar energy is also partly responsible for the generation of oceanic waves and currents as well as regulating the salt/fresh water mix through the melting and re-freezing of the polar ice caps. This supports a conducive living environment for ocean creatures as the oceanic currents circle and deposit shared nutrients across the globe.

The melting of the ice by solar energy makes sure that new fresh water is made available for countless ecosystems around the world that are dependent upon this cycle for survival.

Top Reasons to Use Solar Electricity instead of Grid

1. **Clean**

    Solar energy is clean, renewable, and good for the environment. By using renewable solar energy to meet a portion of your household's electric needs, you can significantly reduce your household's contribution to the release of pollutants.

2. **Reliable**

   Sun will rise every day and when sun rise we get solar energy and we can harness it in all days even though the intensity will vary with location, clouds, weather variations etc. However it is the most reliable source of power. It can be predicted with reasonable accuracy depending on the location with historical data. You can plan a backup system for days with insufficient solar intensity.

3. **Easy to harness**

   We need only stationary panels with no moving parts for harnessing solar energy. Considering how other forms of energy is generated, solar power generation is extremely simple for a host of applications. Once installed, there is little maintenance, the solar panels will continue to generate power without interruption.

4. **Abundant**

   Solar power is abundant with no shortage whatsoever. The energy required by the whole world for one year is received by earth from sun in half an hour.

5. **Good history**

   The history of solar power usage is full of interesting facts. There has been good success whenever human race have tried to harness solar energy.

6. **Affordable**

   Using the sun's free energy to heat water helps lower your home energy bills. Water heating is the second highest energy cost in a typical household. Depending on how much hot water you use, and your current water heating fuel, these savings can be hundreds of rupees a year. If you have a swimming pool, solar is the most economical way to heat your pool and extend your swimming season. A solar pool heating system can pay for itself in four years or less.

7. **Independent power**

   Renewable energy systems provide your home or business with increased independence. Reducing dependence on fossil fuel sources provides long-term protection from growing energy costs and uncertain supplies.

8. **Its nature's power**
   Solar energy is totally natural, nothing artificial about it. As any product of nature, solar power is also simple and gel well with the ecosystem
9. **Is safe to use**
   Solar power is under normal circumstances safe. Only extreme exposure may be dangerous. Otherwise there are neither serious hazards nor side effects related with solar power.
10. **Commitment to our future generations—sustainable living**
    Using available solar power is one of the best ways of sustainable living. Sun delivers power every day, whether we effectively use it or not. The beauty is in using this wonderful power rather than generate power by excavation and pollute the environment.
11. **Harnessing equipment is modular**
    Solar panels are modular. It can be added more when your requirement has increased. Similar case for batteries. This helps in planning the size of equipment and investing according to the need.
12. **Harnessing equipment is removable and reinstalable**
    In case the necessity arises the solar power system equipments can be moved from one place to another. There are even a few sects of nomads who carry a compact solar system on the back of their animal and use it during night wherever they say. If planned well, the solar power system can be taken along when there is a transfer or change of house.
13. **Emergency power**
    Solar electric systems can provide your household with emergency back-up electricity in the case of storm caused or other utility outages. At these times solar electric power systems can work along with a conventional generator or alone. They can also be designed to provide power to critical loads over extended periods of cloudy weather.
14. **Remote power**
    If you plan to build away from established utility service, you should consider the cost of installing a utility line needed to provide the utility's energy. Often, the cost of extending conventional power to your residence is more expensive than the solar option. Solar electric systems are also often a good choice

for providing electricity for use in areas that don't have convenient power sources near by.

15. **Educational power**
By investing in a renewable energy system you provide a hands-on demonstration of clean power. You can teach your friends, neighbors, and family about how energy systems work, and about how energy choices impact our environment.

16. **Easy to troubleshoot**
With hardly 4 main components, the solar power system can be troubleshooted without much difficulty. A few days of systematic training can empower the common man. The training should be done under experienced technicians and supervised well.

17. **Country needs extra power from renewables**
It is a reality now that country needs extra power for its growth apart from that is available from fossil products. Solar is the best option and can supplement the power requirement.

18. **Government is supporting solar**
The government of India realising the potential of solar power has rolled out several measures to promote its usage. This has created a good environment to buy and install solar power system.

19. **Solar energy in inexhaustible**
Sun has been providing energy to the earth for centuries now and will continue to provide energy for many centuries to come. The entire energy falling on earth can be harnessed forever.

20. **No location constraint**
Solar power can be harnessed wherever there is sunshine with no major constraint in location. Even though a few hilly regions are exceptions, mostly solar can be harnessed everywhere. Further the intensity can be predicted with decent accuracy.

Advantages and Disadvantages of Solar Energy

Solar energy is a reliable and never ending source of energy. Although the initial cost of using the solar energy may be high, the high price can be recovered as after the system is installed, there will be no cost of producing the electricity.

Advantages of Solar Energy

- Solar cells are long lasting sources of energy which can be used almost anywhere.
- They are particularly useful in areas where there is no national grid and in areas where there is less population, such as in a remote site.
- Solar cells provide cost effective solutions to energy issues in places where there is no mains electricity.
- Solar cells are totally silent and non-polluting.
- As they have no moving parts, they require little or no maintenance at all, and have a long lifetime.
- Solar cells offer more advantages over other renewable sources; as wind and water power relies on turbines which are noisy, expensive and more liable to breaking down.
- Rooftop power is a good way of supplying energy to a fast growing community.
- More cells can be added to homes and businesses as the community grows so that energy generation is in line with demand.
- Solar cells can also be installed in a distributed fashion, i.e. they don't need large-scale installations.
- Solar cells can easily be installed on roofs, which mean no new space is needed and each user can quietly generate their own energy.

Disadvantages of Solar Energy

- The initial setup cost is quite high.
- Most types of solar cells require large areas of land to achieve a good range of output.
- Air pollution and weather can also have a large effect on the efficiency of the cells.
- The silicon used in cells is also very expensive.
- Currently, solar energy costs higher than traditional sources i.e. coal, oil etc.

Solar Technologies to mitigate climate change:

1. **Displace fossil fuel used for heating / cooling etc**
   Fossil fuels are been used with high consumption for heating, cooling etc throughout the world. This on an hourly basis is negatively impacting the climate and atmosphere. By using Solar energy on an incremental basis, we can make the heating, cooling etc more environmental friendly.
2. **Displace fossil fuel used for electricity generation**
   The exhaustive deployment of solar power harnessing can reduce the dependence on fossil fuel for electricity generation. This is now increasingly evident with the deployment of solar power plants throughout the world.
3. **Displace petroleum transportation fossil fuel use**
   Fossil fuels are excavated at one place, refined at another and put to use at a different place. This involves huge transportation mechanisms and all these vehicles consume fossil fuel. This consumption can be minimized with use of solar energy available through out the planet.
4. **Reduce transmission losses**
   Fossil fuel other power plants are located far away from the place of actual use. Transmission of power brings in huge losses. This can be avoided if the power generation happens nearby. With Solar now it is a reality. Many, but not all, PV applications are distributed, meaning that they are installed at or very near the point of use of the electricity generated. Hence, there is no need for transmission lines, and distribution lines are either short or unnecessary.
5. **Lower or eliminate energy demand in buildings**
   Commercial buildings account for about 18% of the total primary energy consumption in the United States. Story is similar in many countries. Currently a lot of energy is used for lighting during the day in buildings around the world. By proper daylight design, this can be avoided. This can be augmented with the installation of Solar PV system on the roof top of buildings which are now mainly left unused. In the zero-energy building (ZEB) concept, the essence is a building designed in such a way that on-site energy consumption is reduced to a level that can be

met entirely by on-site renewable energy production over a typical one-year period. Here solar power forms an integral part.

6. **Reduce the need for other building components by using BIPV**
   The elegant components of Building Integrated PV system can replace other materials used for roofing, walls etc. The fossil fuel used for manufacturing these roofing, wall materials would have negatively impacted the environment, whereas the BIPV starts generating power on its own the time it is installed and help reduce the consumption of fossil fuel.

7. **Feed power to hybrid /electric vehicles and reduce pollution**
   Burning of fossil fuel by passenger vehicles is a serious contributor to pollution. This can be reduced if more vehicles run on electricity / hybrids are put to use. The charging stations powered by solar panels eliminate the use of fossil fuel for running these pollution free vehicles.

8. **Improve energy efficiency and thereby use less power**
   Use of energy efficient loads is a prime requisite for using solar power. When we use energy efficient load, the total energy consumption come down helps to reduce the consumption from the grid when required.

9. **Avoid hazardous gas and CO2 emission from cooking**
   Burning derivatives of fossil fuel and firewood for cooking emit hazardous gas and CO2. This can be reduced with the use of solar power for cooking. Solar cooking is now in its nascent stage. But if planned judiciously, solar cooking can be put to use for every community based cooking or cooking at individual home.

10. **Generate hydrogen by hydrolyzing water for fuel cells**
    Hydrogen for use in a fuel cell, whether for use to power a vehicle or provide power for a home or business, can be produced by reforming a hydrocarbon. But hydrogen can also be produced by hydrolyzing water, which requires electricity in the process of splitting hydrogen and oxygen from the water. As in the example above, if the electricity source is PV and CSP, then the hydrolysis process is a sustainable, clean process. The use of PV-generated electricity to produce hydrogen is also a storage option for this technology: if the electricity is not needed for another purpose while the sun is shining, it can then be used to store energy by

producing hydrogen for later use, perhaps at night, or other times when there is little available sunlight.

Understanding the above ten applications will help us have a fair commitment on usage of solar power.

An interesting case study:

Two friends Ajay and Abi decide to build a house in the plot that they purchased earlier. These plots are in a housing colony adjacent to each other where they had over 300 days of good sunlight. (In India many locations are blessed with this)

Both went to an architect and made detailed plan about the house— Two floors each of 1000 sqft with car park, water sump, overhead tank etc. Now an estimate was done considering full completion including the floor tiling, wardrobes, furniture, modular styled kitchen, lights and cabling etc to fix a budget of 50 lacs INR. Once this was done the architect suggested the idea of using solar for the house. First reaction from Ajay and Abi was solar is costly, it will not work, cannot use when we want etc. But architect tried to explain the benefits and gave a few literature on solar to both of them. Both went home unconvinced about solar.

After a couple of days, an article about the promotion of solar by the government was published in the news paper. Ajay happened to read this. He was curious and tossed the idea of the architect to his wife who took a neutral stand. Ajay met Abi and showed the paper clipping, but Abi was not interested. Ajay by now was more curious and wanted to know more. He called up the architect and asked from where he could get more details about solar. Architect readily gave the contact number of a company selling solar in the town. Ajay now got in touch with the company and had a few rounds of discussion and then understood

- His entire house can be powered by solar
- He should install a system that will generate 4units of power every day and battery bank with 3 days of autonomy
- It will cost him about 7lakhs to get this done now and 1 lac once in every five years
- He will need to keep about 200 sqft facing south in the roof to install the solar panel

He was also convinced that

- He will not have any power cut / disruptions for next 15 years
- He will not experience any high voltage surge coming from EB line
- He need not install a transformer near his house nor he need to dig up and lay cables on the road
- He need not pay the security deposit to the electricity board
- He need not pay the bill every month

Ajay was clearer that he will be using the cleanest power available to power his dream home. He went and explained all this to his wife who now seconded the idea of going solar. Then he met Abi and shared about solar. But Abi was negative. By now Ajay was convinced that he will make serious effort to install solar, but will finalise after discussing with the architect. Ajay met the architect and explained his wish of installing solar within the total budget of 50 lacs. Architect was receptive and readily agreed to work on the cost of other materials / work to fund 7 lacs for solar within the total budget. Ajay was okay to work on the flooring tiles, furniture, kitchen setup etc and they both could find the fund for solar after 3 to 4 sittings and finalized the budget at 52lacs. They also finalized on having good daylighting for all the rooms by even using lightpipes wherever required. The concept of rainwater harvesting was also discussed and the underground water sump was redesigned to store more water.

Ajay also finalized the total power load for a family of five people to be 1200 VA for Refrigerator, Washing machine, Computer, TV, Music system, microwave, Iron Box, Water pump, Fan Lights, Mixi etc with all water heating for bathing/ washing vessels to be done by solar water heater.

Since the budget and electric power was finalized, Ajay gave a go ahead to the architect once the funds were arranged. He was proud that he was building a lovely home self sufficient with electric power and rain water. The architect on his part finalised the design in 4 weeks and completed the house in 20 weeks ready to move in.

Abi has applied for electric connection and after several documentation and inspection by officials, the transformer was finalized. The road need to be dug up and wired. Since there was minimum planning of loads, Abi ended up having a total load of 2000 VA for the same size of family as Ajay's. The electricity permit et all took 12 weeks and then architect was given a green signal to go ahead. The design was finalized in 4 weeks and the building work started and completed in 36 weeks due to intermittent delay in getting electric connection. Further the total delay brought in a price escalation of 20%.

Now both the friends started living next to each other.Ajay's home became a model for modern living. Abi was paying electricity bill of 3 to 4 k every month with power cuts while electricity of Ajay was free and no power interruptions.

Ajay was saving large amount of carbondioxide emissions every year while the primary energy consumption for electricity to Abi was emitting large amount of green house gases. Ajay with no extra cost has now built a home for sustainable living.

I am sure even if 10% of the capable people here adopt the Ajay way, India will be in the forefront of protecting mother earth and its ecosystems. **It's long over due.**

10 ways to Move off the grid:

The term "Move off the Grid" in its idealistic meaning is simply shutting off the grid connection and using solar/ other renewable energy for the entire power consumption of our home. However if it necessitates it can be done in a phased manner. A few important steps are mentioned below.

1. Switch off lights when not needed
2. Use energy efficient lights and appliances only
3. Do not overload refrigerator, Buy correct size and star rated
4. Use energy efficient fan only. Use Air conditioners only when it is an absolute must
5. Switch off TV and computer fully rather than keeping in stand by mode.
6. Use good quality cables and electrical accessories to avoid losses
7. Design home / office with good day lighting
8. Use solar water heater for heating water. Do not use electrical or gas heaters
9. Install a power back up AC/DC solar system for use during power shut down
10. Install a full AC solar PV system to power the entire house. **Keep grid connection in stand by.**

Read the bill:
Electricity bill payment is normally a monthly activity. It is very important to correctly read the electricity bill. Many among us will look at the final payable amount only. But knowing the energy consumed for the month will give a better understanding of the household power requirement. The number of units consumed has to be monitored every month. By working on this in a systematic manner through better usage of the household loads and engaging Renewable energy sources, consumption of electricity from the grid can be reduced.

16.5 Crore households in India currently consume about 170 Tera WHr of electricity every year and this corresponds to about 2.7 units per household per day. This consumption is growing at a rate of over 15% per annum.

If a reasonable percentage of households adopt Renewables/ Solar and save on an average at least 1 unit of energy every day the impact will be huge. Considering the highest requirement of electricity in future will come from domestic households justify this. Most important to remember is 1unit saved in consumption is 2units saved in supply i.e. 1 unit saved from one customer can be used for another customer and the total supply requirement will come down by 2 units. And for every 2 units saved, over 5000 units of primary energy consumption and related emission will be reduced annually by every household.

Undoubtedly this will give a significant boost for our effort to build a better planet.

CHAPTER 11

HOW TO SIZE A SHS & SWH

> *"I have no doubt that we will be successful in harnessing the sun's energy. If sunbeams were weapons of war, we would have had solar energy centuries ago."*
>
> —*George Porter*
> *Nobel Prize winner in Chemistry, 1967*

Solar home systems—PV electricity:

Major system components

As seen earlier the major components for solar PV system are solar panel, solar charge controller, inverter, battery bank, auxiliary energy sources and loads (appliances). These components should be selected according to your system type, site location and applications.

- **PV module**—Converts sunlight to DC electricity
- **Solar Charge Controller**—regulates the voltage and current coming from the PV panels going to battery and prevents battery overcharging and prolongs the battery life
- **Inverter**—Converts DC output of PV panels /wind turbine or battery into a clean AC current for AC appliances or to feed back to grid
- **Battery**—stores energy for supplying to electrical appliances when there is a demand
- **Load**—is electrical appliances that are connected to solar PV system such as lights, radio, TV, computer, refrigerator etc

Solar PV system sizing

A. Solar Module sizing

The first step in designing a solar PV system is to find out the total power and energy consumption of all electrical loads that need to be supplied by the solar PV system from the Energy Estimation Table.

Energy Estimation Table

Sl No	Appliance	Rated Power (w)	Qty	Total Power reqd (w)	No of hrs of usage	Total Energy required (wh)
1	LED bulb	2	2	4	2	8
2	CFL bulb	7	2	14	2	28
3		7	3	21	3	63
4	CFL bulb	9	2	18	2	36
5		9	1	9	3	27
6	CFL bulb	11	2	22	2	44
7		11	3	33	3	99
8	Tube Light	40	1	40	3	120
9	Fan	50	3	150	8	1200
10	Television	120	1	120	4	480
11	Mixer	300	1	300	0.25	75
12	Computer	200	1	200	1	200
	Total	**766**		**931**		**2380**

PV module Capacity	850 Wp
Inverter Capacity	1400 VA
Charge Controller:	Short circuit current of PV modules x 1.3
Battery Capacity	600 Ah

A.1 Calculate total Watt-hours per day for each appliance
Multiply the power (watts) of individual appliance with its qty and hours of usage every day

A.2 Calculate total Watt-hours per day needed from the PV modules.
Add the total appliances Watt-hours per day to get the total Watt-hours per day which must be provided by the panels.

A.3 Calculate the total Watt-peak rating needed for PV modules
Divide the total Watt-hours per day needed from the PV modules by 2.8 to get the total Watt-peak (Wp) rating needed for the PV panels needed to operate the appliances.

A.4 Calculate the number of PV panels for the system
Divide the total Wp required by the rated output Watt-peak of the PV modules available to you. Increase any fractional part of

result to the next highest full number and that will be the number of PV modules required.

A.5 Calculate the roof space required to mount the PV panels

The PV panel size may vary with manufacturer. Especially the north south and east west wise. As a base line, the space required for crystalline panels will be One sq ft for 10 to 12 Wp. For Thin film the space required will be 40 to 60% more.

B. Inverter sizing

An inverter is used in the system where AC power output is needed. The input rating of the inverter should never be lower than the total watt of appliances. Always prefer to install a pure sine wave high efficiency inverter for solar applications. The inverter must have the same nominal voltage as your battery. The inverter must be large enough to handle the total amount of Watts you will be using at one time. The inverter size should be 15-20% higher than total Watts of appliances. In case of appliance type is motor or compressor then inverter size should be minimum 3 times the capacity of those appliances and must be added to the inverter capacity to handle surge current during starting. Alternately an inverter that will withstand 3 times the current for a brief period will suffice.

C. Battery sizing

As written earlier the battery type recommended for using in solar PV system is deep cycle battery. Deep cycle battery is specifically designed to be cyclically discharged to low energy level and then recharged day after day for years. The battery should be large enough to store sufficient energy to operate the appliances at night and cloudy days. The size of battery can be calculated as follows:

C.1 Calculate total Watt-hours per day used by appliances.
C.2 Divide the total Watt-hours per day used by 0.85 for battery loss.
C.3 Divide the answer obtained in item C.2 by 0.6 for depth of discharge.
C.4 Divide the answer obtained in item C.3 by the nominal battery voltage
C.5 Multiply the answer obtained in item C.4 with days of autonomy (the number of days that you need the system to operate

when there is no power produced by PV panels) to get the required Ampere-hour capacity of deep-cycle battery.

Battery Capacity (Ah) =
Total Watt-hours per day used by appliances x Days of autonomy
(0.85 x 0.6 x nominal battery voltage)

Battery Selection

The following are a list of factors that need to be considered while selecting a battery.

- Required days of storage (autonomy)
- Maximum allowable depth of discharge
- Daily depth of discharge requirements
- Temperature and environmental conditions
- Cyclic life and/or calendar life in years
- Maintenance requirements
- Sealed or unsealed
- Cost and warranty.

D. Solar charge controller sizing

The solar charge controller is typically rated against Amperage and Voltage capacities. Select the solar charge controller to match the voltage of PV array and batteries and then identify which type of solar charge controller is right for your application. Make sure that solar charge controller has enough capacity to handle the current from PV array.

For the series charge controller type, the sizing of controller depends on the total PV input current which is delivered to the controller and also depends on PV panel configuration (series or parallel configuration).

As a standard practice, the sizing of solar charge controller is to take the short circuit current (Isc) of the PV array, and multiply it by 1.3

Solar charge controller rating = Total short circuit current of PV array x 1.3

Note: For MPPT charge controller sizing will be different.

Charge Controller Selection

The following are a list of factors that need to be considered while selecting a charge controller.

- System voltage
- PV array and load currents
- Battery type and size
- Regulation and load disconnect set points
- Mechanical design and packaging
- System indicators, alarms, and meters
- Over current, disconnects and surge protection devices
- Costs, warranty and availability

Important Note:
While selecting any components of the Solar Home System, always pick the next higher capacity one. This will ensure uninterrupted supply of solar energy to your stated requirement.

Earthing:

Proper earthing /grounding of your electrical system are essential to the equipments and your safety. Electricity always follows the path of least resistance, and that path could be your costly equipment or you whenever proper earthing is not done. This becomes more relevant here since the solar home system components including the high efficiency inverter will risk damage if the circuit is not properly earthed.

Grounding directs electrical energy into the earth by providing a conductor that is least resistant. This is accomplished by attaching one end of the wire to the frame of an appliance and fastening the other end to grounding wire. A good grounding/ earthing system protects the solar home system components from lightning to a large extent. Indian Electricity Rules 1956, National Electrical Code, IS 3043 etc provide good inputs to earthing standards and maintenance.

Regular maintenance of the earthing system is a must for its effective performance. Since the conductivity of the earthing wire, plate and the ground around it will have seasonal variations, adhering to a scheduled maintenance will give good results.

Main Components and key characteristics for building a good AC solar system:

Solar Home System—1.5 Units/ Day

Solar Panel:

- Rated Module Power (3 hrs peak sunshine)—535 Wp
- Shadow Free Area Required　　　　—4.5 Sq mtr
- Size (Wp) x No of panels　　　　　—130x3+65x1+50x1+30x1
- Total weight of the Panels　　　　　—53 Kg

Battery Bank

- Capacity (3 days autonomy)　　　　—365 AH
- Operating Voltage　　　　　　　　—24 V
- Rating　　　　　　　　　　　　　—C20
- Weight of the battery bank　　　　　—220 Kg

Inverter

- Capacity　　　　　　　　　　　　—1 KVA
- Efficiency at rated loads　　　　　　—> 90%
- Efficiency at low loads　　　　　　　—>85%
- Waveform　　　　　　　　　　　　—Pure Sine wave

Charge Controller

- Capacity　　　　　　　　　　　　—30A

Solar Home System—3 Units/ Day

Solar Panel:

- Rated Module Power(3 hrs peak sunshine) —1070 Wp
- Shadow Free Area Required　　　　—9 Sq mtr
- Size (Wp) x No of panels　　　　　—130x8+30x1
- Weight of the Panels　　　　　　　—100 Kg

Battery

- Capacity (3 days autonomy)   —735 AH
- Operating Voltage   —24 V
- Rating   —C20
- Weight of the battery bank   —490 Kg

Inverter

- Capacity   —1.5 KVA
- Efficiency at rated loads   —> 90%
- Efficiency at low loads   —>85%
- Waveform   —Pure Sine wave

Charge Controller

- Capacity   —58A

Solar Home System—5 Units/ Day

Solar Panel:

- Rated Module Power(3 hrs peak sunshine)   —1785 Wp
- Shadow Free Area Required   —14.2 Sq mtr
- Size (Wp) x No of panels   —130x12+85x2
- Weight of the Panels   —159 kg

Battery

- Capacity(3 days autonomy)   —1225 AH
- Operating Voltage   —24 V
- Rating   —C20
- Weight of the battery bank   —900 Kg

Inverter

- Capacity   —2 KVA
- Efficiency at rated loads   —> 90%

- Efficiency at low loads —>85%
- Waveform —Pure Sine wave

Charge Controller

- Capacity —95A

Solar Home System—8 Units/ Day

Solar Panel:

- Rated Module Power(3 hrs peak sunshine) —2855 Wp
- Shadow Free Area Required —24 sq mtr
- Size (Wp) x No of panels —130x22
- Weight of the Panels —264 Kg

Battery

- Capacity (3 days autonomy) —980 AH
- Operating Voltage —48 V
- Rating —C20
- Weight of the battery bank —850 Kg

Inverter

- Capacity —2.5 KVA
- Efficiency at rated loads —> 90%
- Efficiency at low loads —>85%
- Waveform —Pure Sine wave

Charge Controller

- Capacity —77A

Solar Home System—10Units/ Day

Solar Panel:

- Rated Module Power(3 hrs peak sunshine) —3570 Wp
- Shadow Free Area Required —29.5 Sq mtr
- Size (Wp) x No of panels —130x27+65x1
- Weight of the Panels —331 Kg

Battery

- Capacity (3 days autonomy) —1225 AH
- Operating Voltage —48 V
- Rating —C20
- Weight of the battery bank —1100 Kg

Inverter

- Capacity —3 KVA
- Efficiency at rated loads —> 90%
- Efficiency at low loads —> 85%
- Waveform —Pure Sine wave

Charge Controller

- Capacity —95A

The above figures are indicative only. The actual requirement will vary according to the conditions at site, usage and make. You should check with your local supplier and finalise the actual configuration.

Solar water Heaters:

It is extremely important to get the sizing done correctly of the solar water heater system. The solar water heater sizing needs to be done based on the hot water requirements and the hot water use habits of the people in a family. The basic idea of having a solar water heater is to reduce electricity consumption for water heating. An under-sized system is insufficient to meet the hot water requirement, an over sized system will result in overheating of the water. As back-up system is required for cloudy days, it may be possible to manage with marginal back up use in extreme weather and optimize the size of the system for use in the rest of the year.

The requirement of hot water varies from person to person. However, it is estimated that the average hot water requirement per person per day for bathing in an average household in India is around 40 liters at 40°C.

The size of the solar water heater should be finalized after considering

1. The hot water requirement of the household taking the usage pattern into consideration
2. The qty of hot water the system can supply

1) Hot water requirement calculation.

In a household the suitable hot water temperature for a male adult is 38 to 40°C, for a female adult is 36 to 38°C and for children it is 34 to 36°C and for infant it is 32 to 34°C. The above is a general pattern considering the softness of the skin and may vary from person to person. However for our calculation here, 40°C can be taken as the requirement for residential applications

Bathing per person per day 40 ltrs
Wash basin per person per day 5 ltrs
Kitchen wash per person per day 5 ltrs
Clothes wash per person per day 10 ltrs
Heat loss and hot water wastage factor 10%

Based on the above table, we can calculate the approximate size required for solar water heating system.

Hot water calculation for an average household

Serial No	Description	Hot water reqd per person	No of persons	Household requirement (ltrs)
1	Bathing	40	2	80
2	Wash basin	5	4	20
3	Kitchen wash	5	4	20
4	Cloth washing	10	2	20

5	Others	5	2	10
	Total requirement			150
	Wastage / Efficiency loss		10%	15
	Usable hot water to be generated			**165**

2) It should be noted that the heat available to get hot water per day is limited and the net output of hot water from the solar water heater will depend on the ambient temperature during day and night and the temperature of the cold water fed to the solar water heater.

HOCI flow:

The solar water heaters used in India predominantly work on HOCI flow concept.

HOCI is Hot water Output Cold water Input

—this means the solar water heater takes in cold water as much as hot water is drawn from the system. This result in mixing of hot and cold water inside the storage tank. This process inside the storage tank and its effect need to be clearly understood before working on the sizing of a Solar water heater.

The performance of the solar water heater is also influenced by

- The time difference from the first drawn hot water ltr to the last drawn ltr on a day
- The temperature of the cold water input
- Using automatic mixer tap or manual mixing with separate taps

Solar hot water output will vary from place to place where there is major difference in ambient temperature which has a direct effect on the temperature of the cold water temperature Places with colder ambient temperature will need higher sized solar water heater.

Solar water heater can generate about 55 to 65°C of hot water. Since our usage will be at **40°C**, the size of the solar water heater needs to match this.

Sl No	Place	Solar water heater capacity required against usable water at 40°C			
		Usable: 150 Ltrs @40 °C	Usable: 200 Ltrs @ 40°C	Usable: 300 Ltrs @ 40°C	Usable: 400 Ltrs @ 40°C
1	Srinagar	89	118	178	237
2	Amristar	88	117	175	233
3	New Delhi	90	120	180	240
4	Lucknow	82	109	164	218
5	Kolkotta	77	102	154	205
6	Ahmedbad	75	100	150	200
7	Mumbai	69	92	138	184
8	Bhopal	75	100	150	200
9	Hyderabad	75	100	150	200
10	Bengaluru	79	105	157	210
11	Chennai	71	95	142	189
12	Kochi	71	95	142	189
13	Patna	82	109	164	218
14	Pune	79	105	157	210
15	Mangalore	67	89	133	178

The above list has tabulated the capacity of solar water heater required at different cities (having varying ambient temperature) to deliver the qty of hot water as per the header row.

After arriving at the quantity of the hot water generation and storage, the size of the SWH should be decided in discussion with the vendor in your town. The generation capacity will change according to the collector area provided and the solar irradiance at the particular location during different seasons.

Important Note:
Generally the Solar Electric home systems and Solar water heaters will have a life span that stretches your requirement well beyond your present

one. Hence adequate capacity should be built into the calculation to support your requirement for a longer period. Please ensure you contact a reputed solar products seller/ installer to get the correct configuration and sizing to suite your need.

CHAPTER 12

UAE GREEN CITY

> *"We cherish our environment because it is an integral part of our country, our history and our heritage. On land and in the sea, our forefathers lived and survived in this environment. They were able to do so because they recognized the need to conserve it, to take from it only what they needed to live and to preserve it to the succeeding generations"*
>
> *Sheikh Zayeed Bin Sultan Al Nahyan*

In Abu Dhabi, construction is on for the world's first carbon-neutral city. This amazing concept is expected to blossom as a silicon valley of clean technologies thereby creating an enticing environment for innovation and business in this sector. Masdar means "source" in Arabic.

With the completion of Masdar Initiative, Abu Dhabi will position itself to play a vital role in supporting renewable energy business around the world.

Masdar City is to be constructed on an area of approximately 6 square kilometers, located about 17 kilometers east of the capital Abu Dhabi. It is designed to support a population of about 50,000. It is being designed by British architects Foster and Partners and it will cost between £10bn ($15bn) and £20bn ($30bn).

Initiated in 2006, the project is projected to take about eight years to build. Construction began in early 2008; by 2016, Masdar will be what is billed as world's most ambitious green business project—a zero-emission city with a clean-tech university and the foundations laid to become the world's clean-tech Mecca. By establishing a center for clean-tech engineers and financiers to collaborate, Masdar will be a hub of creativity and a source for new ideas and technologies that will seriously impact the demand and supply for world's future energy.

The project is headed by the Abu Dhabi Future Energy Company (ADFEC). The city is planned to cover 1,500 businesses, primarily commercial and manufacturing facilities specialising in environment friendly products, and more than 60,000 workers are expected to commute to the city daily. Automobiles will be banned within the city; travel will be accomplished via public mass transit and personal rapid transit systems, with existing road and railways connecting to other locations outside the city. The absence of motor vehicles coupled with Masdar's perimeter wall, designed to keep out the hot desert winds, allows for narrow and shaded streets that help funnel cooler breezes across the city.

Masdar City will be the latest of the highly planned, specialized, research and technology-intensive municipalities that incorporate a living environment, similar to grand boarding university.

Background of Masdar:

> "A new era is upon us, challenging us to venture beyond the achievements of the past and meet the needs of the future"
> —Ahmed Ali Al Sagyeh (Chairman)

> "Abu Dhabi is evolving its global energy leadership through the Masdar initiative—demonstrating long term commitment to renewable energy for a sustainable future"
> —Dr. Sultan Ahmed Al Jaber (CEO)

Established in 2006, Masdar is a commercially driven enterprise that operates to reach the broad boundaries of the renewable energy and sustainable technologies industry—there by giving it the necessary scope to meet these challenges.

Masdar operates through five integrated units, including an independent, research-driven graduate university, and seeks to become a leader in making renewable energy a real, viable business and Abu Dhabi a global centre of excellence in the renewable energy and clean technology category. The result is an organisation greater than the sum of its parts and one where the synergies of shared knowledge and technological advancement provide this commercial and results-driven company with a competitive advantage that includes an ability to move with agility and intelligence within an industry that is evolving at great speed.

This holistic approach keeps Masdar at the forefront of this important global industry, while ensuring it remains grounded in the pursuit of pioneering technologies and systems that also are feasible. As a result, it delivers innovation to the market while deriving profits for its shareholders.

Masdar is a wholly owned subsidiary of the Abu Dhabi Government-owned Mubadala Development Company, a catalyst for the economic diversification of the Emirate.

Masdar City

Aspiring to be one of the most sustainable cities in the world, approximately 6km² Masdar City is an emerging global clean-technology cluster that places its resident companies in the heart of the global renewable energy and cleantech industry. Situated 17km from downtown Abu Dhabi, Masdar City is a high-density, pedestrian-friendly development where current and future renewable energy and clean technologies are showcased, marketed, researched, developed, tested and implemented.

The city, which at full build out will house 40,000 residents and hundreds of businesses, will integrate the full range of renewable energy and sustainability technologies, across a living and working community. As with most dynamic technology clusters, the city has a top-notch research university that is a source for innovation, technologies, R&D and highly skilled graduates.

Masdar Power

Masdar Power is a developer and operator of renewable power generation projects. In building a portfolio of strategic utility-scale projects, Masdar Power makes direct investments in individual projects in all areas of renewable energy, with a focus on Concentrating Solar Power (CSP), photovoltaic solar energy and on—and offshore wind energy. In a joint venture with Abengoa Solar and Total, Masdar Power is developing the 100MW Shams 1 CSP plant in the Western Region of Abu Dhabi, set to be the largest CSP plant in the world.

It also is developing a 30MW wind farm and a PV array on Sir Bani Yas Island. Through these and future projects, the unit will contribute to Abu Dhabi's goal of generating 7% of its energy needs from renewable sources. International projects include the 1,000 MW London Array offshore wind farm and a wind farm in the Thames Estuary, a joint venture with DONG Energy an E.ON, that when completed will be the largest offshore wind farm in the world. It also is building a wind farm in the Seychelles that will provide 25% of the island's energy needs.

Masdar Power also strategically invests in technology relevant to utility-scale renewable energy. Companies in which it holds significant ownership stakes include Torresol, a joint venture with SENER Grupo de Ingeniería of Spain to build and operate Concentrating Solar Power (CSP) plants globally and Masdar PV, a thin-film solar panel manufacturer in Germany.

Masdar Capital

Masdar Capital seeks to build a portfolio of the world's most promising renewable energy and clean technology companies. It helps its portfolio companies grow and scale-up by providing capital and management expertise. Masdar Capital targets investments that have the greatest potential globally and to the UAE and is particularly focused on the following sectors:

- **Clean energy:** including power generation and storage technologies, transportation technologies, cleantech /clean energy innovation, and sustainable biofuels.
- **Environmental resources:** including water and waste management, and sustainable agriculture technologies.
- **Energy and material efficiency:** including developments in advanced materials, building and power-grid efficiency, and the enabling technologies.
- **Environmental services:** including environmental protection and business services.
  Investment in these markets is made via two funds: the Masdar Clean Technology Fund (MCTF), launched in 2006, and the DB Masdar Clean Tech Fund (DBMCTF), launched in 2009. MCTF, a fully deployed $250 million fund invested $45 million

in three cleantech funds and the remaining $205 million in 12 direct investments in companies, as lead or co-lead investor. It was launched in conjunction with partners Consensus Business Group, Credit Suisse and Siemens AG. DBMCTF, is jointly managed with Deutche Bank and raised US$265 million in its first close, has an initial investor group led by Siemens and includes the Japan Bank for International Cooperation, Japan Oil Development Co. Ltd., Nippon Oil Corporation, Development Bank of Japan and GE.

Both funds follow an active management investment strategy. The targeted investment amount is between US$15-35 million and seeks to realise strong risk-adjusted returns. Through these funds,
Masdar Capital also seeks to demonstrate, commercialise and promote renewable technologies in the UAE, and to identify synergies between its investments and other Masdar activities, as well as the long-term energy and development programme of the UAE.

Masdar Carbon

Masdar Carbon manages projects that bring reductions in carbon emissions through energy efficiency and waste heat/CO2 recovery, as well as through carbon capture and storage (CCS). The unit provides value to industrial asset owners by monetising carbon emission reductions under the current United Nations-based Clean Development Mechanism (CDM) or other applicable future international climate trading schemes, and by providing an end-to-end solution to achieve this, including carbon finance, project identification and management, technology sourcing project analysis and registration at the United Nations. Masdar Carbon's geographic focus under the CDM is the Middle East, Africa and Asia, while its sector focus is on oil, gas and power. Masdar's CDM project portfolio includes a diversified range of projects focusing on gas flaring reduction, gas leakage reduction, combined heat and power, industrial CO2 recovery and solar power.

Masdar Carbon also has entered into a joint venture with E.ON Carbon Sourcing to invest in carbon abatement projects in Africa, the Middle East, and Central and Southeast Asia. The company will develop, finance and implement projects in the Middle East, Africa and Asia with a particular focus on power generation and oil and gas. The emission reductions will be monetised in the form of carbon credits and traded under the CDM or other applicable future_schemes.

As part of its mandate to invest in technologies for the production of clean fossil fuels, Masdar Carbon is developing one of the world's most ambitious large-scale CCS projects in partnership with the Abu Dhabi National Oil Company (ADNOC), its group of companies, and others in the power and industrial sectors in the Emirate. The project will capture carbon dioxide emitted from power plants and heavy industry and transport it, via a national pipeline network, for injection into Abu Dhabi's oil & gas reservoirs for enhanced oil recovery. The first phase of the project is currently in the front-end engineering and design stage and upon completion, will capture five million tons of carbon dioxide per year. Masdar Carbon will contribute to Plan Abu Dhabi 2030 by helping to lower the Emirate's carbon footprint.

Masdar Institute

The Masdar Institute is an independent, research-driven graduate institute developed with the ongoing support and cooperation of the Massachusetts Institute of Technology (MIT). Other major partners include Siemens, which will establish its Middle East headquarters and Centre of Excellence in Building Technologies R&D centre; GE, which will build its first ecomagination Centre; Schneider, which will operate an R&D centre; BASF; the Swiss Village Association; the Korea Technopark Association, and the International Renewable Energy Agency (IRENA).

Focused on the science and engineering of advanced alternative energy, environmental technologies and sustainability, the Masdar Institute will be at the heart of the home-grown research and development community at Masdar City and will eventually host 600-800 Master's and PhD students and 200 faculty.

The graduate programmes integrate education, research and scholarly activities to prepare graduate students to be innovators, creative scientists, researchers and critical thinkers in the areas of technology development, systems integration and policy. Masdar Institute partners with industry and government to foster a diversified knowledge-based economy in Abu Dhabi and the UAE.

As a crucial source of research and development, the institute is fundamental to Masdar's core objectives of developing Abu Dhabi's knowledge economy and finding solutions to humanity's toughest challenges. The university aims to become one of the world's leading academic institutions in its field. The high quality of its students and faculty, as well as its unique location within Masdar City, will enable it to achieve these goals.

Renewable Resources:
Masdar will employ a variety of renewable power resources. Among the first construction projects will be a 40 to 60 megawatt solar power plant, built by the German firm Conergy, which will supply power for all other construction activity. This will later be followed by a larger facility, and additional photovoltaic modules will be placed on rooftops to provide supplemental solar energy totalling 130 megawatts. Wind farms will be established outside the city's perimeter capable of producing up to 20 megawatts, and the city intends to utilise geothermal power as well. In addition, Masdar plans to host the world's largest hydrogen power plant.

Water management has been planned in an environmentally sound manner as well. A solar-powered desalination plant will be used to provide the city's water needs, which is stated to be 60 percent lower than similarly sized communities. Approximately 80 percent of the water used will be recycled and waste water will be reused "as many times as possible," with this grey water being used for crop irrigation and other purposes.

The city will also attempt to reduce waste to zero. Biological waste will be used to create nutrient-rich soil and fertilizer, and some may also be utilised through waste incineration as an additional power source. Industrial waste, such as plastics and metals, will be recycled or re-purposed for other uses.

On June 21, 2009, the Fraunhofer Gesellschaft and the Abu Dhabi Future Energy Company, representing the Masdar City Project, signed a cooperative agreement for a strategic partnership. Over the long term the goal is to establish a close cooperation in the field of sustainable urban development and building planning. Participating in the cooperation are the Fraunhofer Institutes for Industrial Engineering IAO and for Building Physics IBP as well as the Fraunhofer Institute for Solar Energy Systems ISE.

Of the many research areas, an important focus will be on developing more efficient ways of producing, storing, and transmitting wind and solar power, as well as hydrogen-based fuels.

Other work areas will be: energy efficiency, carbon capture and storage (CCS), and water management and desalination—critical in the dry Middle East.

Absent from Masdar will be the sound and smell of cars hooting' and spewing'. Conventional vehicles will be parked outside of the walled boundary, but people weary of walking this compact and pedestrian-friendly city can use pod cars below street level. These are equipped with magnetic sensors that detect potential obstacles and are powered by the sun.

All the energy used in Masdar will be renewably generated, not only the electrical power, but also that for heating, cooling and transport. The bulk of this is likely to come from solar of one form or another. There will be power generation for a smart grid from solar thermal power and concentrating PV, and also distributed PV throughout the city. The wind resource in Abu Dhabi is generally poor and will contribute little to the overall mix, but some geothermal and waste-to-energy, particularly from biowaste, are also likely to be significant contributors.

Solar and other forms of energy

The main source of the city's energy supply will be solar energy, be it rooftop installations or concentrated solar troughs, from two huge solar power plants under construction.

A recent German Aerospace Center's study shows that the greatest energy potential of the Middle East lies in solar power, as there is no lack of sunlight in the region. However, there has also been investment in wind power; and the Abu Dhabi government is examining other renewable energy alternatives.

These plans include thermal tubes to be integrated with the city's buildings to provide hot water, and possibly a deep geothermal "hot rock" borehole to provide a constant source for a 24-hour cooling system.

At night, though, Masdar City will have to import gas-fired energy from Abu Dhabi. But Masdar will still be overall carbon neutral, as the solar power plants will generate so much excess energy during the sunny hours of the day that it can be exported to Abu Dhabi.

A Clean-tech Silicon Valley—tax-free

By creating an academic foundation, Masdar planners hope to establish a synergy between science and business, and to incentivize clean-tech companies worldwide to build branches and partnerships in Masdar.

According to Dr. Sultan Ahmed Al Jaber, CEO of the Masdar Initiative, a fundamental goal of the project is to overcome the fragmented global approach to research, development, and marketing with clean technologies. "There are other examples of pioneering alternative energy ventures," he says, "but they are under-funded, under-supported by regulators, and under—recognized. Each country does not have to reinvent the wheel"

But a new green Silicon Valley does not form overnight, so the Masdar planners have also decided to entice entrepreneurs by promising a tax-free business environment, a high degree of intellectual property protection, and as few or as transparent of regulations as possible.

Further, by concentrating clean-tech R&D, Masdar is intended to function as a financing machine to spur the growth of green inventions. This Masdar project is one of the aggressive approaches to support renewables. Let us hope the commitment of the team ensure that it is completed on time with all the stated objectives. Once such a city is established which is carbon neutral, zero emission, low noise, zero waste etc—it is an ideal project which can be emulated at other places / countries that are capable and serious about the future of our planet.

CHAPTER 13

Energy Efficiency

> *"We generate our own environment. How can we resent a life we've created ourselves? Who's to blame, who's to credit but us? Who can change it, anytime we wish, but us?"*
> —*Richard Bach*

The energy efficiency is perhaps one of the most important issues facing the human race today. The efficient use of energy was advocated since the time we started using energy conversion devices / equipments for our modern day living. But the concept of energy efficiency was more or less discarded every time a new fuel / technology were discovered. This apart energy efficiency closely followed the growth of power generation. The surging demand of power all round the world, depletion of fossil fuels, climate change impact and above all the increasing cost of power and related economic and social issues have now brought energy efficiency as a mainstay in decision making of any developmental projects.

The two important aspect of energy efficiency are (1) Using a highly efficient device / equipment (2) Running a device / equipment to its best possible efficiency.

These two can be seen our daily life through

(1) Using a highly efficient device / equipment:

   (a) Using LED as against CFL as against Incandescent bulbs
   (b) Designing and constructing an energy efficient building

(c) Buying and using a fuel efficient vehicle
(d) Using an energy efficient cooking appliances
(e) Selecting a better efficient computer/ Laptop

(2) Running a device/ equipment efficiently

(a) Use lights and fans when only required
(b) Ensure correct loading of refrigerator and washing machine
(c) Run a car in its highest fuel efficient speed
(d) Use good automatic level controller for water pumping system and use water without wastage
(e) Use sunlight and open window breeze during daytime rather than switching electric lights and fans

The resources are limited and it is the fact—the fuel for transportation or to provide power for all by which the modern world operates are not as endless as was once believed. Natural gas and fossil fuel sources that have been relied on so heavily for over a century are being depleted at rapidly increasing rates, due to the high demands of the developed nations, surging demand from the developing nations and increase in appetite of under developed countries.

The technologies and their applications that are inefficient use energy in excess and continue to release high levels of harmful elements that negatively affect air, water, and land resources, further degrading the environment's ability to recharge and sustain itself.

Energy efficiency has an immediate impact all round and requires fully integrated and proactive public policies from National, state level and municipal programs. Once enacted and executed in an effective way less money and effort will be needed to address the problems of pollution and resource sustainability, thereby freeing these resources for other productive use.

Energy Conversion Efficiency:

"Energy can neither be created nor destroyed. It can only be converted from one form to another"—This is one of the underlying principle that stress the importance of energy efficiency.

Energy conversion efficiency is the ratio of useful output of an energy conversion machine to the energy input. The useful output may be electric power, mechanical work, or heat. Energy conversion efficiency is not defined uniquely, but instead depends

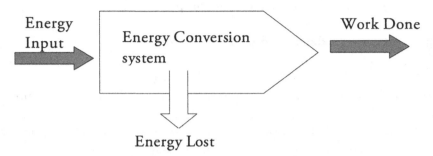

Output energy is always lower than input energy

on the usefulness of the output.

Generally, energy conversion efficiency is a dimensionless number between 0 and 1.0, or 0 to 100%.

More specific terms include

- Electrical efficiency, useful power output per electrical power consumed;
- Mechanical efficiency, where one form of mechanical energy (e.g. potential energy of water) is converted to mechanical energy (work);
- Thermal efficiency or Fuel efficiency, useful heat and/or work output per input energy such as the fuel consumed;
- 'Total efficiency', e.g., for cogeneration, useful electric power and heat output per fuel energy consumed. Same as the thermal efficiency.
- Luminous efficiency, that portion of the emitted electromagnetic radiation is usable for human vision.

Efficient energy use, sometimes simply called energy efficiency, is the goal of efforts to reduce the amount of energy required to provide products and services. For example, insulating a home allows a building to use less heating and cooling energy to achieve and maintain a comfortable temperature. Installing fluorescent lights or natural skylights reduces the amount of energy required to attain the same level of illumination compared to using traditional incandescent light bulbs. Compact fluorescent lights use two-thirds less energy and may last 6 to 10 times longer than incandescent lights.

In many countries energy efficiency is seen to have a national security benefit because it can be used to reduce the level of energy imports from foreign countries and may slow down the rate at which domestic energy resources are depleted.

Energy productivity:

Energy efficiency is a cost-effective strategy for building economies without growing energy consumption. In the seventies physicist Amory Lovins popularized the notion of a "soft energy path", with a strong focus on energy efficiency. Among other things, Lovins popularized the notion of negawatts—the idea of meeting energy needs by increasing efficiency instead of increasing energy production.

Energy productivity is the output and quality of goods and services per unit of energy input. Energy productivity can be improved from either reducing the amount of energy required to produce something, or from increasing the quantity or quality of goods and services from the same amount of energy.

Making homes, vehicles, and businesses more energy efficient is a largely untapped solution to address the problems of pollution, global warming, energy security, and fossil fuel depletion.

Appliances:

Modern energy-efficient appliances such as refrigerators, freezers, ovens, stoves, dishwashers, and clothes washers and dryers, use significantly less energy than older appliances. For example current energy efficient refrigerators use 30 to 40 percent less energy than conventional models did 15 years back. There are a large number of appliances in our country that were installed long back when energy efficiency was not keenly monitored. Hence these appliances use a large amount of energy and deliver lesser output compared to their latest counterparts. If all these appliances are replaced by new ones, significant energy can be saved.

Many studies have shown that the replacement of old appliances is one of the most efficient global measures to reduce emissions of greenhouse gases. Modern power management systems also reduce energy usage by idle appliances by turning them off or putting them into a low-energy mode after a certain time. Many countries identify energy-efficient appliances using energy input labeling.

Building Design:
A building's location and surroundings play a key role in regulating its temperature and illumination. For example, trees, landscaping, and hills can provide shade and block wind. In cooler climates, designing buildings with a south facing windows increases the amount of sun (ultimately heat energy) entering the building, minimizing energy use, by maximizing passive solar heating. Tight building design, including energy-efficient windows, well-sealed doors, and additional thermal insulation of walls, basement slabs, and foundations can reduce heat loss by 25 to 50 percent.

Proper placement of windows and skylights and use of architectural features that reflect light into a building can reduce the need for artificial lighting. Increased use of natural and task lighting have been shown by one study to increase productivity in schools and offices. Newer fluorescent lights produce a natural light, and in most applications they are cost effective, despite their higher initial cost, with payback periods as low as a few months.

Effective energy-efficient building design can include the use of latest techniques of low cost Passive Infra Reds (PIRs) to switch-off lighting when areas are unoccupied such as toilets, corridors or even office areas out-of-hours. In addition, lux levels can be monitored using daylight sensors linked to the building's lighting scheme to switch on/off or dim the lighting to pre-defined levels to take into account the natural light and thus reduce consumption. Building Management Systems (BMS) link all of this together in one centralised computer to control the whole building's lighting and power requirements.

A lot of work is being done in the field of Net Zero Energy buildings. Most net zero energy buildings harvest energy through a combination of solar and wind while using the latest techniques to reduce the consumption of energy. Net zero energy buildings are also known as Zero carbon buildings or Zero emissions building.

Industry:
The energy intensive sectors include fertilizers, Aluminum, iron and steel, cement etc and these use very large amount of energy to power a diverse range of manufacturing and resource extraction processes.

Industrial processes require large amounts of heat and mechanical power, most of which is delivered as natural gas, petroleum fuels and as electricity. In addition some industries generate fuel from waste products that can be used to provide additional energy.

A lot of possible opportunities are available to improve energy efficiency in industry. Many depend on the specific technologies and processes in use at each industrial facility. Various industries generate steam and electricity for subsequent use within their facilities. When electricity is generated, the heat that is produced as a by-product can be captured and used for process steam, heating or other industrial purposes. Conventional electricity generation is about 30 percent efficient, whereas combined heat and power (also called co-generation) converts up to 90 percent of the fuel into usable energy.

Advanced boilers and furnaces can operate at higher temperatures while burning less fuel. Using Solar heat for / as a pre feed will significantly reduce the use of energy. These technologies are more efficient and produce fewer pollutants.

Vehicles:

There are several ways to enhance a vehicle's energy efficiency.
A few are

- Improved aerodynamics
- Reduced weight
- Tyres to reduce friction
- Inflated tyres rather than deflated one
- Timely cleaning of air filters
- Computer controlled engines
- Using right gear at the right speed
- Proper usage of accelerator and clutch
- Good road condition
- Judicious use of Car AC

Energy-efficient vehicles may reach twice the fuel efficiency of the average automobile. Cutting-edge designs, such as the diesel Mercedes-Benz Bionic concept vehicle have achieved a fuel efficiency as high as 84 miles per US gallon (2.8 L/100 km; 101 mpg_{-imp}), four times the current conventional automotive average.

The mainstream trend in automotive efficiency is the rise of electric vehicles. Plug-in hybrids, Hybrids, and all electric vehicles improve efficiency. They recapture energy through regenerative braking which would have dissipated in normal cars, especially pronounced for city driving. . Plug-ins can typically drive for around 40 miles (64 km) or more purely on electricity without recharging; if the battery runs low, a gas engine kicks in allowing for extended range.

Energy Conservation:
Energy conservation is much wider than energy efficiency. It includes active efforts to reduce the consumption through scheduling and planning. In home or office we should schedule our work and usage of right size equipments to ensure energy consumption is restricted. This may involve behavioral change, deployment of the right size machines/equipments, training on usage, preventive maintenance for upkeep etc. The usage condition can be reduced to reduce the power consumption. You can keep the set temperature of AC to higher level, keep the temperature of room heater to lower level and use public transport instead of own vehicle. In all these cases energy conservation happens.

Efficient energy use and energy conservation are important on environmental and economic terms. This becomes more relevant when we can save fossil fuels. Energy conservation is a challenge requiring policy programmes, technological development and behavioral change. Many organizations

Governmental or non-governmental organisations on local, regional, or national level, are working on often publicly funded programmes or projects to meet this challenge.

Sustainable Energy:
Energy efficiency and renewable energy are said to be the "twin pillars" of a sustainable energy policy. These must be developed concurrently in order to stabilize and reduce carbon dioxide emissions and carbon content. Efficient energy use can slow down the increase in requirement of fossil fuels there by reducing the negative impacts of climate change. Slowing demand growth will only begin to reduce total carbon emissions; a reduction in the carbon content of energy sources is also needed. For this clean energy solutions should start delivering adequate power and capture major share in total energy supply. A

sustainable energy economy thus requires major commitments to both efficiency and renewables.

Rebound Effect:

Rebound effect in simple terms is the increase in total energy consumption even though energy efficiency is implemented. Under a good energy efficiency program, there will be saving in energy and hence the energy requirement should reduce. But this will not be the case always. Many a times when the requirement of energy is reduced through energy efficiency measures, the period of usage of the service increases; thereby the total energy consumption also increases. This is rebounding of energy usage and hence called rebound effect. An example will be people using the car with more mileage to drive more distance.

This is a serious subject and will be predominantly seen when we consider energy efficiency only on economic terms. Consumption of cheaper services will increase and the total consumption of the resource will escalate. If not understood and controlled on time, the entire benefit of implementing energy efficiency will be washed out by this rebound effect. Hence energy efficiency should be implemented with specific focus on social and environmental impact along with economic benefits.

Estimates of the size of the rebound effect range from roughly 10% to 30%. In some cases it may be even more. Since more efficient (and hence cheaper) energy will also lead to faster economic growth, it is expected that improvements in energy efficiency unless controlled may eventually lead to even faster resource use.

The Energy Conservation Act 2001

The energy conservation act was enacted by Govt of India in March 2002 to address the energy conservation primarily through improving energy efficiency with active support form central and state governments. It is expected among other things empower the governments to

- specify energy consumption standards for notified equipment and appliances;
- direct mandatory display of label on notified equipment and appliances;
- prohibit manufacture, sale, purchase and import of notified equipment and appliances not conforming to energy consumption standards;

- notify energy intensive industries, other establishments, and commercial buildings as designated consumers;
- direct every owners or occupier of a new commercial building or building complex being a designated consumer to comply with the provisions of energy conservation building codes;
- direct designated consumers to comply with energy consumption norms and standards
- Establish Bureau of Energy Efficiency

and lot more. This act was amended in 2010 to give it more ammunition to address the implementation of energy conservation.

Bureau of Energy Efficiency (BEE)

The Government of India set up Bureau of Energy Efficiency (BEE) (Website: http://www.bee-india.nic.in) on 1st March 2002 under the provisions of the Energy Conservation Act, 2001. The mission of the Bureau of Energy Efficiency is to assist in developing policies and strategies with a thrust on self-regulation and market principles, within the overall framework of the Energy Conservation Act, 2001 with the primary objective of reducing energy intensity of the Indian economy. This will be achieved with active participation of all stakeholders, resulting in accelerated and sustained adoption of energy efficiency in all sectors.

Role of BEE

BEE co-ordinates with designated consumers, designated agencies and other organizations and recognize, identify and utilize the existing resources and infrastructure, in performing the functions assigned to it under the Energy Conservation Act. The Energy Conservation Act provides for regulatory and promotional functions.

The Major Regulatory Functions of BEE include:

- Develop minimum energy performance standards and labeling design for equipment and appliances
- Develop specific Energy Conservation Building Codes
- Activities focusing on designated consumers
- Develop specific energy consumption norms
- Certify Energy Managers and Energy Auditors
- Accredit Energy Auditors

- Define the manner and periodicity of mandatory energy audits
- Develop reporting formats on energy consumption and action taken on the recommendations of the energy auditors

The Major Promotional Functions of BEE include:

- Create awareness and disseminate information on energy efficiency and conservation
- Arrange and organize training of personnel and specialists in the techniques for efficient use of energy and its conservation
- Strengthen consultancy services in the field of energy conservation
- Promote research and development
- Develop testing and certification procedures and promote testing facilities
- Formulate and facilitate implementation of pilot projects and demonstration projects
- Promote use of energy efficient processes, equipment, devices and systems
- Take steps to encourage preferential treatment for use of energy efficient equipment or appliances
- Promote innovative financing of energy efficiency projects
- Give financial assistance to institutions for promoting efficient use of energy and its conservation
- Prepare educational curriculum on efficient use of energy and its conservation
- Implement international co-operation programmes relating to efficient use of energy and its conservation

Mission:

The mission of Bureau of Energy Efficiency is to "institutionalize" energy efficiency services, enable delivery mechanisms in the country and provide leadership to energy efficiency in all sectors of the country. The primary objective would be to reduce energy intensity in the economy.

The broad objectives of BEE are as under:

- To exert leadership and provide policy recommendation and direction to national energy conservation and efficiency efforts and programs.

- To coordinate energy efficiency and conservation policies and programs and take it to the stakeholders
- To establish systems and procedures to measure, monitor and verify energy efficiency results in individual sectors as well as at a macro level.
- To leverage multi-lateral and bi-lateral and private sector support in implementation of Energy Conservation Act and efficient use of energy and its conservation programs.
- To demonstrate delivery of energy efficiency services as mandated in the EC bill through private-public partnerships.
- To interpret, plan and manage energy conservation programs as envisaged in the Energy Conservation Act.

Objectives:

- Provide a policy recommendation and direction to national energy conservation activities
- Coordinate policies and programmes on efficient use of energy with shareholders
- Establish systems and procedures to verify, measure and monitor Energy Efficiency (EE) improvements
- Leverage multilateral, bilateral and private sector support to implement the EC Act 2001
- Demonstrate EE delivery systems through public-private partnerships

The Bureau would obtain inputs and co-opt expertise from private sector, non-governmental organisations, research institutions and technical agencies, both national and international, to achieve these objectives.

National Energy Conservation Award:
Ministry of Power, through BEE, organizes the annual energy conservation awards function on the occasion of National Energy Conservation Day on the 14th December. These awards recognize innovation and achievements in energy conservation by the Industry; Commercial Buildings, Railways and help raise awareness about the need and efficacy of energy conservation and efficiency

National Mission for Enhanced Energy Efficiency:
This national mission is part of the 8 missions formulated under the nation Action plan on climate change. Draft outlined in 2008 and financial outlay approved by the cabinet in 2010, this mission focus on improving the energy efficiency in India and thereby saving huge amount of power.

The four initiatives to drive this mission are

- Perform Achieve and Trade(PAT)—The market based system to enhance the cost effectiveness in improving the energy efficiency in energy intensive industries through certification of energy savings that can be traded
- Market Transformation for Energy Efficiency (MTEE)—Accelerating the shift of energy efficient appliances in designated sectors through innovative measures to make products more affordable
- Energy Efficiency Financial Platform (EEFP)—creation of mechanisms that would help finance demand side management programmes in all sectors by capturing future energy savings
- Framework for Energy Efficient Economic Development (FEEED)—Develop fiscal instruments to develop energy efficiency.

The mission is aiming to save 23 MTOE and mitigate 98 million tones of CO_2 emission. The infrastructure is being put in place and is expected to gain momentum soon with Power ministry and BEE driving it.

World Energy Council—Indian Member committee (WEC—IMC):

Vision:
To be the most effective member committee truly representative of Indian energy sector and contributing to furtherance of energy goals of the country.

Mission Statement:

- Facilitate review, research and advocacy of energy sector technology, policy & strategy

- Provide a platform for authoritative sharing of information on Indian energy sector
- Synergize relevant information with member committees worldwide for ensuring long term sustainability of supply & use of energy

WEC's work is categorized into four main areas:

- **Global Studies**, which focus on current or emerging high visibility issue facing the global energy sector
- **Technical Programmes**, which address ongoing performance, technology or policy issues and involve regular collection, analysis and dissemination of specialized energy information and data as well as benchmarking, best practices and standards.
- **Regional Programmes**, which focus on specific regional issues and priorities, taking the form of regional forums, workshops, regionally focused initiatives and facilitation of regional "signature projects"
- **Communications and Outreach**, which seeks to increase WEC's visibility and influence and deliver its message to a global audience.

Wesite: http://indiaworldenergy.org

Alliance for an Energy efficient economy (AEEE):

Alliance for an Energy Efficient Economy (AEEE) is a member-driven industry association providing a common platform for energy efficiency (EE) stakeholders to collaborate and address barriers to energy efficiency in India, through policy research, facilitating market transformation, fostering technology innovations, capacity building of energy professionals and stimulating financial investments.

The following are AEEE's Focus Areas:

- **Agenda driven by Members:** AEEE mandate is driven by its members towards promoting energy efficiency (EE).
- **Policy Research and Advocacy:** Programmes and activities are developed in response to members' priorities in their bid to participate in EE markets and policy making set forth by BEE and Government of India.

- **Capacity building in M&V and Professional Certification:** In response to industry demand AEEE organizes M&V training programmes on the International the Performance & Verification Protocol (IPMVP), developed and conducted by Efficiency Valuation Organization (EVO).
- **AEEE Platform for Collaboration:** AEEE conducts roundtables and conferences to facilitate member interactions with BEE and collaborations with various national and international organizations for facilitating market transformation and for fostering EE technology innovations.
- **Promotion of Demand Side Management:** AEEE is working on developing best practices and models for promoting energy efficiency and Demand Side Management.
- **SME Expertise and Networks:** AEEE includes distinguished members whose activities with SMEs have won them awards for entrepreneurship in micro and small enterprises.
- **AEEE's partnership with other Associations:** AEEE is a Member Association of Confederation of Indian Industry (CII), and a Member of CII-Indian Green Building Council (IGBC).

Website: www.aeee.in

Source: Respective Websites

Tips for energy efficiency in our daily life

Lighting System

1. Natural daylight is the best light available. Use open curtains, shades, skylights wherever possible
2. LED lights are the most energy efficient. CFL bulbs come next. The incandescent bulbs consume excess energy should be the least preferred.
3. One of the best energy-saving devices is the light switch. Turn off lights when not required.
4. Employing infrared sensors, motion sensors, automatic timers, dimmers etc help in saving energy used in lighting.
5. Plan your room lighting. Be specific where you need whole room lighting and where you need task lights.

6. Clean your tube lights and lamps regularly.
7. Control outdoor lighting by using dusk to dawn controller.

Room Air Conditioners

1. Use energy efficient ceiling or table fan wherever adequate.
2. You can reduce air-conditioning energy use by as much as 40 percent by shading your home's windows and walls. Plant trees and shrubs to keep the day's hottest sun off your house.
3. ACs will consume 3 to 5 percent less energy for each degree set above 22°C, so set the thermostat of room air conditioner at 25°C to provide the most comfort at the least cost.
4. A Ceiling or room fan allows better air movement which in turn helps cool the room faster.
5. A good air conditioner will cool and dehumidify a room in about 30 minutes, so use a timer and leave the unit off for some time.
6. Always keep doors to air-conditioned rooms closed.
7. Clean the air-conditioner filter every month. A dirty air filter reduces airflow and may damage the unit. Clean filters enable the unit to cool down quickly and use less energy.
8. If room air conditioner is older and needs repair, it's likely to be very inefficient. It may work out cheaper on life cycle costing to buy a new energy-efficient air conditioner.

Refrigerators

1. Make sure that refrigerator is kept away from all sources of heat including direct sunlight, radiators and appliances such as the oven, and cooking range.
2. Ensure the door gasket of the refrigerator is in good condition to avoid heat seepage.
3. Refrigerator condenser coils and compressors dissipate heat, so allow enough space for continuous airflow around refrigerator.
4. Allow adequate air circulation inside a refrigerator as this improves the efficiency.
5. Think about what you need before opening refrigerator door. You'll reduce the amount of time the door remains open.

6. Allow hot and warm foods to cool and cover them well before putting them in refrigerator. Refrigerator will use less energy and condensation will reduce.
7. Clean the condenser coils regularly to make sure that air can circulate freely.
8. For manual defrost refrigerator, accumulation of ice reduces the cooling power by acting as unwanted insulation. Defrost freezer compartment regularly for a manual defrost refrigerator.

Water Heater

1. Always insulate hot water pipes, especially where they run through unheated areas.
2. By reducing the temperature setting of water heater from 60 degrees to 50 degrees C, one could save over 18 percent of the energy used at the higher setting.
3. Use aerating low flow faucets and shower heads
4. Never allow any leaks in the hot water circuit
5. Buy the right sized electric water heaters. Heating excess water will not only take time, but also burn unwanted energy

Microwave Ovens

1. Microwaves save over 40% of energy by reducing the time to cook. Especially for small qty of food.
2. Microwave oven cook food from the outside edge toward the centre of the dish, so if you're cooking more than one item, place larger and thicker items on the outside.
3. Choose the right size to suit your need. Using a large sized one for cooking small quantity of food will be an inefficient process.
4. Keep the over rack clear. Do not foil the racks and use pans that allow adequate air flow.
5. Ensure the microwave ovens you buy have temperature probes, controls to turn off when food is cooked, variable power setting etc that will help boost energy efficiency.

Electric Kettles

1. Use an electric kettle to heat water. It's more energy efficient than using an electric cook top element.
2. When buying a new electric kettle, choose one that has an automatic shut-off button and a heat-resistant handle.
3. A clean kettle is a must for efficient cooking. Hence regularly clean the kettle to remove mineral deposits.
4. Use kettle to heat the required qty for one time use. Do not fill and heat water required for the whole day.
5. Ensure you are attentive always when you use the kettle. Do not leave it unattended while heating the water.

Computers

1. Turn off your home office equipment when not in use. A computer that runs 24 hours a day consume more power than an energy-efficient refrigerator.
2. If your computer must be left on, turn off the monitor; this device alone uses more than half the system's energy.
3. Setting computers, monitors, and copiers to use sleep-mode when not in use helps cut energy costs by approximately 40%.
4. Using LCD monitors instead of conventional CRT monitor help save energy.
5. Reduce the brightness of the monitor to the optimum level. Higher brightness does not mean higher visibility, but only higher energy consumption.

Automobile

1. Drive in right gear: Always drive the vehicle in the right gear. Driving in a lower gear than you need waste fuel. Driving a vehicle in top gear on hills and on corners also waste fuel.
2. Drive Smoothly: Drive vehicle as smooth as you can. Intermittent brakes and gear changes, clutch release all will end up in partial burning of fuel. Hence plan your drive through roads and time your drive when you have lower traffic
3. Avoid over speeding: Many engine manufacturers indicate 40 to 50 km/hr as the speed of most fuel efficiency. Read your manual

carefully and adhere to the recommended speed for best fuel efficiency. Higher speeds will use more fuel and deliver less power.
4. Upkeep your tyres and air filters: Keep your tyres inflated to the required level and clean your air filters periodically. This is more important when we drive on our challenging roads and polluted environment.
5. Keep your vehicle in good condition: Always ensure periodic maintenance is done as scheduled in the owner's manual. Right engine tuning, proper fuel circuit, balanced wheel, taunt belt drives are all a prerequisite for the engine to perform efficiently.

Cooking efficiently

1. Cover the vessels while cooking. Ensure Blue flame when cooking with LPG
2. The bottom of the vessel should cover the flame/ coil completely. An exposed flame/ coil will lose heat
3. Keep the correct amount of gas flow/ electric supply and cut off the supply just before the food is cooked to use the residual heat
4. Use the correct amount of liquid. Excess water use will prolong cooking time
5. Always prefer to cook using the pressure cooker. This will help reduce the energy required.

Laundering efficiently

1. Use lower temperature setting and proper detergents. Utilize the cold and warm water wash cycles effectively
2. Segregate the clothes while drying. Thinner clothes need lesser time to dry, while thicker clothes will need more time to dry
3. Dry full loads as possible and do not add wet clothes intermittently
4. Use outside drying as much as possible
5. Set the correct program for the qty of clothes and save water and energy.

Electric Motor / Pump

1. Use the correct size of the pump. Read the manual of the pump before selecting and get correct advice
2. Use the right size of plumbing line items. Using a ½ inch pipe with a 3/4th inch pump will seriously dampen the output
3. Locate the pump at the nearest position of the water source
4. Ensure correct voltage supply and proper electrical connection/ contact
5. Maintain and upkeep the pump as suggested by the manufacturer

Electric Iron

1. Iron only fully dried clothes. Do not iron when the cloth is wet
2. Use electric iron with auto cut off function and set the correct temperature
3. Do not sprinkle excess water while ironing
4. Switch on the electric iron only when you need to and switch off just before finishing the ironing.
5. Choose the iron with the appropriate power for you as the range is wide from 200wats to 1500 watts.

Television

1. Buy the required size TV. A very large TV in an inadequate room will only reduce the watching experience and only increase the energy bill
2. Turn off the quick start option: The option even though will start the TV a few seconds earlier, but will take excess energy
3. Control the LCD backlight effectively. Excess backlight will not only reduce the viewing experience, but also waste energy
4. Turn the TV off when not required, Do not leave the TV on as a background when you are not really watching it.
5. Switch the TV fully off rather than leaving it in standby mode when you are not watching it for a prolonged period.

Audio Systems

1. Buy the right sized audio equipment for your house
2. Inventorise a good energy efficient one before finalising the equipment
3. Do not leave it switch on when not required and do not put in stand by mode.
4. High wattage speakers use more energy—plan your sound system to get the best effect with good aquastic planning
5. Do not keep audio system on when watching Television

Power Backup Inverter

1. Calculate the load requirement and buy the correct size inverter
2. A more efficient sine wave inverter even though costly is always better than a low efficient one. Inverters with good efficiency at lower loads is a definite plus
3. Always use tubular batteries and ensure its regular check and upkeep.
4. Planning exclusive inverter load or priority load will help reduce wastage and proper use of backup power
5. Avoid overloading the inverter both in terms of peak power and power consumption

CHAPTER 14

OTHER RENEWABLE TECHNOLOGIES

> *"The ultimate test of man's conscience may be his willingness to sacrifice something today for future generations whose words of thanks will not be heard"*
> —*Gaylord Nelson, founder of Earth Day*

1) Rain Water harvesting:

Rainfall and soil water are fundamental parts of our global ecosystems. Availability and quality of water determines ecosystem productivity, both for agricultural and natural systems. There is increasing demand on water resources for development whilst maintaining healthy ecosystems, which put water resources under pressure.

Rainwater harvesting is harvesting and storing water in days of abundance, for use in lean days. It is the process of collecting and storing rain water for later productive use.

Storing of rainwater can be done in two ways;

(i) Storing in an artificial storage—This is more specifically called roof water harvesting focusing on human needs of providing immediate relief from drinking water scarcity,

(ii) In the soil media as groundwater—to provide sustainable relief from water scarcity, addressing the needs of all living classes in nature.

Rainwater harvesting is used all over the world:

- **in households** for drinking, cooking, bathing, cleaning—but also for watering small plots and raising small animals,
- **in institutions** like schools, community and religious centers to satisfy their water needs,
- **in agriculture**, to improve yields substantially and at the same time contribute to combating land degradation or flood damage,
- **in urban areas** where it is used as lower quality water for toilet flushing, laundry or gardening but also mitigates storm water run-off and
- **in industry** it is appreciated for its softness, requiring less efforts for purification.
- Rainwater harvesting has been used in ancient Palestine, Greece and Rome. Around 3rd Century BC., farming communities in Baluchistan and Kutch used it for irrigation. In Ancient Tamil Nadu, India, Rainwater harvesting were done by kings. In the Indus Valley Civilization, Elephanta Caves and Kanheri Caves in Mumbai rainwater harvesting alone has been used to supply in their water requirements.
- Currently in China and Brazil, rooftop rainwater harvesting is being practiced for providing drinking water, domestic water, water for livestock, water for small irrigation and a way to replenish ground water levels.
- In Bermuda and US Virgin Island, the law requires all new construction to include rainwater harvesting adequate for the residents.
- United Kingdom encourages fitting large underground tanks to new-build homes to collect rainwater for flushing toilets, washing clothes, watering the garden, and washing cars. This reduces by 50% the amount of mains water used by the home.
- In Tamil Nadu, India rainwater harvesting was made compulsory for every building to avoid ground water depletion. It proved excellent results within five years and every other state took it as role model. Since the implementation, Chennai saw 50 per cent rise in water level in five years and the water quality significantly improved.
- In Rajasthan, India rainwater harvesting has traditionally been practiced by the people of the Thar Desert. There are many ancient water harvesting systems in Rajasthan, which have now been revived

- Kerala, India,—rain water harvesting is pursued aggressively by a large section of society, which is now the emerging trend in the neighboring state of Karnataka

Benefits of Rainwater Harvesting:

1. Environment friendly and easy approach for water requirements
2. RWH is the ideal solution for all water requirements.
3. Increase in ground water level.
4. Mitigates the effects of drought.
5. Reduces the runoff, which other wise flood storm water drains.
6. Reduces flooding of roads and low-lying areas.
7. Reduced soil erosion.
8. Improves the ground water quality.
9. Low cost and easy to maintain.
10. Reduces water and electricity bills.

Climate change will affect rainfall and increase evaporation, which will put increasing pressures on our ecosystem. Similarly, development by a growing population will affect our ecosystems as we increase our demands for services, including reliable and clean water. Rainwater harvesting will continue to be an adaptation strategy for people living with high rainfall variability, both for domestic supply and to enhance crop, livestock and other forms of agriculture.

2) Daylighting:

Daylighting is the controlled admission of natural light—direct and diffused sunlight into a building. By placing windows, other openings and reflective surfaces at different identified location of buildings more sunlight will get in thereby reducing the need for artificial illumination and energy cost.

Daylighting is a direct process to increase energy efficiency that incorporates varying design philosophies and related technologies. Any good daylight design while providing enough sunlight to an occupied space will ensure elimination of undesirable side effects. It ensures balancing of heat gain and loss, glace control etc. The size of the spacing of the windows, glass selection, the reflectance of interior finish, location of interior partitions, light shading techniques etc are all evaluated and used to ensure proper daylight.

Active daylighting is the process of collecting sunlight by tracking the sun and its location/ movement to increase the efficiency of light collection. Passive daylighting is the process of collecting sunlight and reflecting it deeper into a building by using stationery building components and techniques

Daylight can be adopted to any kind of buildings including commercial office buildings, warehouse, factory, school classrooms, cafeterias, hospitals, libraries, inside ships etc.

A few of daylighting techniques/ components are

- Windows—Windows are the best and convenient way to admit sunlight. Being horizontal or vertical determines the amount of sunlight that come in. They allow large amount of light and need to be protected with suitable glass finish to avoid excessive light and heat. Also complements as an element to supply fresh air.
- Skylights—Generally fixed on the roof and are also called as roof window. This consists of a glass mounted on a mounting frame and then integrated on to the roofing. The sealing will be leak proof. When open like a window it provides fresh air as well. Significantly reduce energy costs.
- Light Shelf—mounted as an extension of the window facing the equator. This light shelf reflects the sunlight on to the ceiling of the room through the window thereby illuminating the room. It is a simple but a very effective way to capture the light.
- Light Pipes—Or Solar tubes as it is called are high performance daylights. These carry sunlight from one point to the other. Takes the form of a pipe or tube. Different technologies such as reflective foil, optical fiber, fluorescence based systems etc are all in use. Consists of two portions—one which collect the outside sunlight and the other that will transmit the collected light.
- Heliostats—Mostly track the sun and will continuously reflect direct sunlight to a target space/ wall/ ceiling. Heliostats are mounted to at a position to bisect the angle between the target and the sun and this is rotated as the sun rotates during day time. Heliostats can transmit high power sunlight and are becoming sophisticated with the use of computers.
- Smart Glass—Also called as smart windows are the ones that can change the light transmission properties when a voltage is applied. Technologies include electro chromic devices, liquid crystal

devices etc, their use can save cost of heating, air conditioning and lighting for a building.

- Hybrid Solar Lighting—is a system that uses the solar lights and channelised sunlight to illuminate the interiors. The fiber bundles are led from exterior/rooftop optical light collectors through small openings or cable and carry the light to where it is needed. The artificial lights along with the channelised sunlight to give the targeted illumination of interiors.

3) Wind Energy:

Wind power is the conversion of wind energy into a useful form of energy, such as using wind turbines to make electricity, wind mills for mechanical power, wind pumps for pumping water or drainage, or sails to propel ships.

How it works

Wind turbines works on the principle opposite that of the fan. Fan use electricity to rotate the wings/ blades and produce airflow (breeze). Wind turbines produce electricity by rotation of wings/ blades from airflow (wind). Wind is the flow of air from one location to another. These moving air particles possess kinetic energy. The energy level will depend on the speed of flow. The wind-electric turbine blades are designed to capture the kinetic energy in wind. When the turbine blades capture wind energy and start moving, they spin a shaft that leads from the hub of the rotor to a generator. The generator turns that rotational energy into electricity.

Since wind speed is not constant, a wind farm's annual energy production is never as much as the sum of the generator nameplate ratings multiplied by the total hours in a year. The ratio of actual productivity in a year to this theoretical maximum is called the capacity factor. Typical capacity factors are 20-40%, with values at the upper end of the range in particularly favourable sites. For example, a 1 MW turbine with a capacity factor of 35% will not produce 8,760 MW·h in a year ($1 \times 24 \times 365$), but only $1 \times 0.35 \times 24 \times 365 = 3,066$ MW·h, averaging to 0.35 MW.

Types / Classification

Horizontal axis

Horizontal-axis wind turbines (HAWT) have the main rotor shaft and electrical generator at the top of a tower, and must be pointed into the wind. Small turbines are pointed by a simple wind vane, while large turbines generally use a wind sensor coupled with a servo motor. Most have a gearbox, which turns the slow rotation of the blades into a quicker rotation that is more suitable to drive an electrical generator.

Vertical-axis wind turbines (VAWTs) have the main rotor shaft arranged vertically. Key advantages of this arrangement are that the turbine does not need to be pointed into the wind to be effective. This is an advantage on sites where the wind direction is highly variable, for example when integrated into buildings. With a vertical axis, the generator and gearbox can be placed near the ground, using a direct drive from the rotor assembly to the ground-based gearbox, hence improving accessibility for maintenance.

Small wind turbines
Small-scale wind turbine produces up to 5 kW of electrical power per hour. Small units often have direct drive generators, direct current output, aero elastic blades, lifetime bearings and use a vane to point into the wind.

This is a typical turbine that can be used along with solar to make a hybrid system. These need a little space only and can be easily mounted on rooftop. Wind turbines have been used for household electricity generation in conjunction with battery storage over many decades in remote areas.

This wind turbine charges a 12V battery to run 12V appliances

Off-grid system users can either adapt to intermittent power or use batteries, photovoltaic or diesel systems to supplement the wind turbine.

Offshore wind:

Offshore wind power refers to the construction of wind farms in bodies of water to generate electricity from wind. Unlike the term typical usage of the term "offshore" in the marine industry, offshore wind power includes inshore water areas such as lakes, sheltered coastal areas etc both fixed bottom and floating bottom turbines are used as offshore machines.

Offshore turbines require different types of bases for stability, according to the depth of water. Turbines are much less accessible when offshore (requiring the use of a service vessel for routine access, and a jack up rig for heavy service such as gearbox replacement), and thus reliability is more important than for an onshore turbine

Current scenario

At the end of 2011, worldwide installed capacity of wind-powered generators was 240gigawatts (GW). Several countries have achieved relatively high levels of wind power penetration (with large governmental subsidies), such as 20% of stationary electricity production in Denmark, 14% in Ireland and Portugal, 15% in Spain, and 9% in Germany .Over 80 countries around the world are using wind power on a commercial basis.

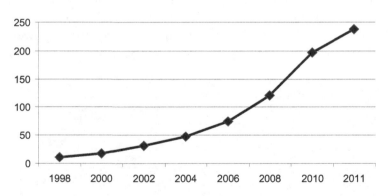

Global Wind Power Installation (GW)

Wind Power installed capacity (GW)—Top ten countries

Rank	Country	2005	2006	2007	2008	2009	2010	2011
1	China	1.3	2.6	5.9	12	26	45	62
2	USA	9	11.5	17	25	35	40	47

3	Germany	18	21	22	24	26	27	29
4	Spain	10	12	15	17	19	21	22
5	India	4.4	6	8	10	11	13	16
6	France	0.8	1.6	2.5	3	4.5	6	6.8
7	Italy	1.7	2	2.7	3.5	5	5.8	6.7
8	UK	1.4	2	2.4	3	4.3	5	6.5
9	Canada	0.7	1.5	1.9	2.4	3.3	4	5.2
10	Portugal	1	1.7	2	2.8	3.5	3.7	4

There are now many thousands of wind turbines operating, with a total name plate capacity of about 250 GW of which wind power in Europe accounts for about 40%, China 26%, USA 20% and balance others. India tops the others with 7% of global installed capacity. World wind installed capacity increased more than 10 times from a meager 17GW to almost 250 GW today. This is astonishing and no power generation system (fossil or non fossil) can match speed of deployment. 50% of wind power installations are in the US and Europe. The share of the top ten countries account for over 85% of total global installations. Of the top ten China, United States, India, United Kingdom and Portugal are experiencing good growth.

Wind accounts for nearly one-fifth of electricity generated in Denmark—the highest percentage of any country.

China is the current leader in installed capacity with over 60GW capacity. It has the unique distinction of a country that could double its wind power installed capacity every year for 5 years (2006 to 2010). Wind power alone generated almost 70 TWh of electricity in China. This growth has brought in the problem of proper distribution of power since the transmission system was not ready for such a surge in power generation. Now things are being streamlined. China is looking towards installing 200GW by 2020.

United States of America currently ranked 2nd has an installed capacity of 47 GW of wind power. It has added substantial amounts of wind power generation capacity, growing from just over 6 GW at the end of 2004 to 8 fold now. Huge thrust in wind power can be seen as Wind power is now the second highest of the new power generation source added Natural Gas being the highest.

India ranks 5th in the world with a total wind power capacity of 16 GW in 2011, or 3% of all electricity produced in India. The wind energy harnessing in India is led by majors like Suzlon, Vestas, Micon among others. India is targeting to reach at least 25 GW by 2017 and considering the wind power potential is over 100GW, this is achievable.

Many other countries around the world are aggressively pursuing wind power generation. These include Mexico, Brazil, South Africa, Canada, and Portugal among others.

Advantages

- Wind energy is friendly to the surrounding environment, as no fossil fuels are burnt to generate electricity from wind energy.
- Wind turbines take up less space than the average power station. Windmills only have to occupy a few square meters for the base; this allows the land around the turbine to be used for many purposes, for example agriculture.
- Newer technologies are making the extraction of wind energy much more efficient. The wind is free, and we are able to cash in on this free source of energy.
- Wind turbines are a great resource to generate energy in remote locations, such as mountain communities and remote countryside. Wind turbines can be a range of different sizes in order to support varying population levels.
- Wind energy when combined with solar electricity, is a great source of energy for developed and developing countries to provide steady, reliable supply of electric power.

Disadvantages

- The main disadvantage regarding wind power is due to the winds unreliability factor. In many areas, the winds strength is too low to support a wind turbine or wind farm, and this is where the use of solar power or geothermal power could be great alternatives.
- Wind turbines generally produce a lot less electricity than the average fossil fuelled power station, requiring multiple wind turbines to be built in order to make an impact.
- Wind turbine construction can be very expensive and costly to surrounding wildlife during the build process.

- Wind turbine when rotating at high speed in turbulent wind will generate high sound and will lead to noise pollution
- Installation of wind turbines may jeopardize the serenity of country side

4) Micro Hydro Energy:

Micro-hydro schemes produce power from streams and small rivers. The power can be used to generate electricity, or to drive machinery. Micro-hydro brings for the first time electricity to remote communities, replacing kerosene for lighting, providing TV and communications to homes and community buildings, and enabling small businesses to start.

Micro-hydro schemes are benefiting many remote communities in the Himalayas and Andes in South America, in hilly parts of China, Sri Lanka and the Philippines. In the developed world, micro-hydro schemes supply power to existing mains electric grids.

How it works:

The power available in a river or stream depends on the rate at

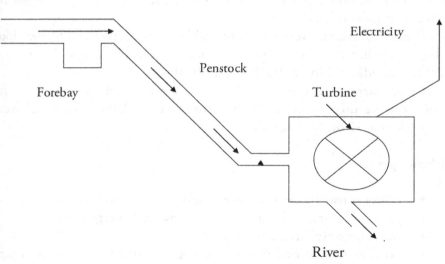

Penstock and turbine house for micro-hydro scheme

which the water is flowing, and the height (head) which it falls down. The core of a micro-hydro scheme is the turbine, which is rotated by the moving water. The turbine rotates a shaft, which is often used to drive an electrical generator.

Hydro schemes are classified by the output power which they produce as approximately:

- Large scale: 2 MW and above
- Mini: 100 kW to 2 MW
- Micro: 5 kW to 100 kW
- Pico: less than 5 kW

Types / Classification

Pelton turbine (for high head, low flow) consists of a set of small buckets arranged around a wheel onto which one or more jets of water are arranged to impact.

Francis turbine (lower head and higher flow) has a spiral casing that directs the water flow through vanes on a rotor.

Cross-flow or Banki turbines (even lower head and higher flow) are made as a series of curved blades fixed between the perimeters of two disks to make a cylinder. The water flows in at one side of the cylinder and out of the other, driving the blades around. They are much easier to make than most other designs.

Propeller turbine (very low head and large flow) has fixed blades, like a boat propeller. A more complex version, the Kaplan turbine, has blades that can be adjusted in pitch relative to the flow.

River current turbine, which is like a wind-turbine immersed in water, can be used to extract power from with a large flow in a river, where there is virtually no head.

Advantages

- power is usually continuously available on demand,
- given a reasonable head, it is a concentrated energy source,
- the energy available is predictable,
- no fuel and limited maintenance are required, so running costs are low (compared with diesel power)
- it is a long-lasting and robust technology; systems can last for 50 years or more without major new investments.
- In remote areas, micro-hydro schemes can bring electricity for the first time to whole communities

- The electrical power from micro-hydro also is sufficient to run machinery and refrigerators, thus supporting small businesses as well as homes

Disadvantages

- it is a site specific technology to sites that are well suited to the harnessing of water power
- there is always a maximum useful power output available from a given hydropower site, which limits the level of expansion of activities which make use of the power,
- river flows often vary considerably with the seasons, especially where there are monsoon-type climates and this can limit the firm power output to quite a small fraction of the possible peak output,
- lack of familiarity with the technology and how to apply it inhibits the exploitation of hydro resources in some areas.

5) Biogas Energy:

Biogas systems use bacteria to break down wet organic matter like animal dung, human sewage or food waste. This produces biogas, which is a mixture of methane and carbon dioxide, and also a semi-solid residue. The biogas is used as a fuel for cooking, lighting or generating electricity. Using biogas can save the labour of gathering and using wood for cooking, minimise harmful smoke in homes, and cut deforestation and greenhouse gas emissions. Biogas plants can also improve sanitation, and the residue is useful as a fertilizer.

Individual biogas systems are already benefiting several million households in Nepal, India, China and elsewhere. Larger systems are also used, for instance to process farm waste in Germany, and at sewage treatment works in the UK.

It is estimated that there are over 2,000,000 small scale biogas plants or digesters in India alone.

How it works

A simple biogas plant has a container to hold the decomposing organic matter and water (slurry), and another to collect the biogas. There must also be systems to feed in the organic matter (the feedstock), to take the gas to where it will be used, and to remove the residue.

A biogas plant needs some methane-producing bacteria to get it started. Once the plant is producing biogas, the bacteria reproduce and keep the process going. Cattle dung contains suitable bacteria, and a small amount of cattle dung is often used as the 'starter' for a biogas plant, even when it is not the main feedstock.

Types / Classification

- **Fixed-dome plants**
- **Floating-drum plants**
- **Balloon plants**
- **Earth Pit plants**

Of these, the two most familiar types in developing countries are the **fixed-dome plants** and the **floating-drum** plants.

Fixed Dome Plant:

A fixed-dome plant consists of a digester with a fixed, non-movable gas holder, which sits on top of the digester. When gas production starts, the slurry is displaced into the compensation tank. Gas pressure increases with the volume of gas stored and the height difference between the slurry level in the digester and the slurry level in the compensation tank.

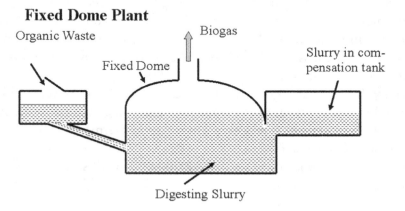

Fixed Dome Plant

The costs of a fixed-dome biogas plant are relatively low. It is simple as no moving parts exist. There are also no rusting steel parts and hence a long life of the plant (20 years or more) can be expected. The plant is constructed underground, protecting it from physical damage and saving space.

Types of fixed-dome plants

- **Chinese fixed-dome plant** is the basic model of all fixed dome plants. Several million have been constructed in China. The digester consists of a cylinder with round bottom and top.
- **Janata model** was the first fixed-dome design in India, as a response to the Chinese fixed dome plant. It is not constructed anymore. The mode of construction lead to cracks in the gasholder—very few of these plants had been gas-tight.
- **Deenbandhu,** the successor of the Janata plant in India, with improved design, was more crack-proof and consumed less building material than the Janata plant. with a hemisphere digester
- **CAMARTEC model** has a simplified structure of a hemispherical dome shell based on a rigid foundation ring only and a calculated joint of fraction, the so-called weak / strong ring. It was developed in the late 80s in Tanzania.

Floating Drum plants:

Floating-drum plants consist of an underground digester and a moving gas-holder. The gas-holder floats either directly on the fermentation slurry or in a water jacket of its own. The gas is collected in the gas drum, which rises or moves down, according to the amount of gas stored. The gas drum is prevented from tilting by a guiding frame. If the drum floats in a water jacket, it cannot get stuck, even in substrate with high solid content.

Floating Drum Plant

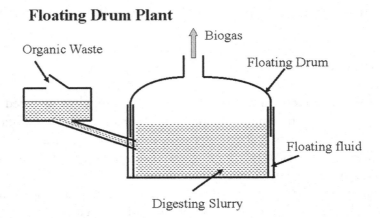

In the past, floating-drum plants were mainly built in India. A floating-drum plant consists of a cylindrical or dome-shaped digester and a moving, floating gas-holder, or drum. The gas-holder floats either directly in the fermenting slurry or in a separate water jacket. The drum in which the biogas collects has an internal and/or external guide frame that provides stability and keeps the drum upright. If biogas is produced, the drum moves up, if gas is consumed, the gas-holder sinks back.

Floating-drum plants are used chiefly for digesting animal and human feces on a continuous-feed mode of operation, i.e. with daily input. They are used most frequently by small—to middle-sized farms (digester size: 5-15m^3) or in institutions and larger agro-industrial estates (digester size: 20-100m^3).

Types of floating drum plants:
There are different types of floating-drum plants:

- **KVIC model** with a cylindrical digester, the oldest and most widespread floating drum biogas plant from India.
- **Pragati model** with a hemisphere digester
- **Ganesh model** made of angular steel and plastic foil
- **Floating-drum plant** made of pre-fabricated reinforced concrete compound units
- **Floating-drum plant** made of fibre-glass reinforced polyester
- **Low cost floating-drum plants** made of plastic water containers or fiberglass drums: ARTI Biogas plants

- **BORDA model:** The BORDA-plant combines the static advantages of hemispherical digester with the process-stability of the floating-drum and the longer life span of a water jacket plant.

Balloon Plant:

A balloon plant consists of a heat-sealed plastic or rubber bag balloon), combining digester and gas-holder. The gas is stored in the upper part of the balloon. The inlet and outlet are attached directly to the skin of the balloon. Gas pressure can be increased by placing weights on the balloon. If the gas pressure exceeds a limit that the balloon can withstand, it may damage the skin. Therefore, safety valves are required. If higher gas pressures are needed, a gas pump is required. Since the material has to be weather—and UV resistant, specially stabilized, reinforced plastic or synthetic materials are given preference. Trevira and butyl. The useful life-span does usually not exceed 2-5 years.

Earth Pit Plant:

These biogas plants are made by a simple earth pit. This will not have thick masonry wall. Instead a thin protective layer is set around the earth pit to avoid seepage. These are build in places with stable soil. The edge of the pit is reinforced with a ring of masonry that also serves as anchorage for the gas-holder. The gas-holder of metal or plastic is located on top of the pit and anchored adequately. The requisite gas pressure is achieved by placing weights on the gas-holder. An overflow point in the peripheral wall serves as the slurry outlet.

Current scenario

Rural families often use animal dung as the feedstock for a biogas plant. The dung from two to four cows (or five to ten pigs) can produce enough gas for all cooking and sometimes lighting too. The family needs to feed the plant once each day with a mixture of dung and water. Ashden Award winner BSP-Nepal coordinates a programme which has sold over 170,000 fixed-dome plants throughout rural Nepal.

Food waste can also be used as the feedstock. Food waste breaks down and produces gas more quickly than dung, so the slurry does not need to be held for as long; these plants are therefore smaller and more suitable for urban homes. A family or community using just their own food waste can replace between 25% and 50% of their cooking fuel. BIOTECH in

Kerala, South India, supplies plants to manage the waste from vegetable markets, and produce gas for electricity generation.

Larger-scale biogas schemes can produce sufficient gas to generate electricity. This is frequently done in sewage treatment plants in the UK, and there are a number of large farm-based plants in Germany and elsewhere. Biogas plants can work well for many years, provided that they are constructed well and checked regularly. If the plant is made from masonry, care must be taken to make sure that the structure is water-tight and gas-tight. The slurry needs to be kept at a temperature of about 35°C for the bacteria to work effectively, and feedstock must be added regularly so that they continue to multiply.

Advantages

- Biogas replaces kerosene or LPG for cooking and cut CO_2 emission
- Biogas cook stoves help reduce the firewood cooking
- Rural household save time since collecting dung and feeding to biogas plant takes much less time than collection firewood and preparing cooking fire
- Biogas plant with stove helps to cook faster during anytime of day and night
- Biogas plant reduces indoor gas pollution
- Help reduce deforestation and thereby related CO_2 concentration in atmosphere.

Disadvantages:

- The process of digestion reduces the total solids content in the sludge and thus there is a volume loss of the organic waste compared to composting.
- Biogas contains contaminant gases which can be corrosive to gas engines and boilers;
- Digestate must meet high standards in order to be used on land without detrimental affects on agricultural uses especially food crops.
- Continuous feed of decomposing matter is required to get good quantity of gas

6) Biomass Energy:

Biomass, a renewable energy source, is biological material from living, or recently living organisms, such as wood, waste, (hydrogen) gas, and alcohol fuels. Biomass is commonly plant matter grown to generate electricity or produce heat. The most conventional way in which biomass is used however still relies on direct incineration. Forest residues, yard clippings, wood chips and garbage are often used for this. Biomass also includes plant or animal matter used for production of fibers or chemicals. Biomass may also include biodegradable wastes that can be burnt as fuel. It excludes organic materials such as fossil fuels which have been transformed by geological processes into substances such as coal or petroleum.

How it works

Chemical Composition:

Biomass is carbon based and composed of organic molecules containing hydrogen and atoms of oxygen. Nitrogen and small quantities of other atoms, including alkali, alkaline earth and heavy metals can be found as well. Plants combine water and carbon dioxide to build sugar blocks. The required energy is produced from light via photosynthesis based on chlorophyll. The sugar building blocks are the starting point for the major energy storage and dissipation during Biomass energy extraction.

Types / Classification

Biomass Sources:
Biomass energy is derived from five distinct energy sources:

- Virgin Wood
- Agricultural Residues
- Energy Crops
- Industrial Waste and co-products
- Food Waste

Virgin Wood: Consists of wood and other products which have no chemical treatments or finishes applied. Virgin wood may be obtained with varying physical and chemical properties from different sources. This include Bark, Brash, Logs, Sawdust, Wood chips, Wood pellets etc.

Agricultural Residue: These residues are both of dry and wet type

Dry type include straw, corn stover etc while wet include animal slurry and farm yard manure

Energy Crops: These crops are grown specifically for fuel and offer high output. There are different types of energy crops suited for different cultivation.

Industrial Waste: These are formed from Industrial processes and manufacturing operations. Untreated wood, Treated wood, Laminates, Paper pulp, Textiles etc all form good source of energy.

Food Waste: These are waste at all points of food supply from initial production, through processing, distribution to post consumer waste from hotels, restaurants etc

Biomass conversion to useful energy:

There are a number of technological options available to make use of a wide variety of biomass types as a renewable energy source. Conversion technologies may release the energy directly, in the form of heat or electricity, or may convert it to another form, such as liquid biofuel or combustible biogas.

Direct Combustion:

Through direct combustion steam or direct heat is produced mainly from residues such as woodchips, sawdust, bark, hog fuel, black liquor, straw, municipal solid waste (MSW), and wastes from the food industry.

Firing the biomass material with fossil fuel is relatively new technique with good advantages for electricity production.

Thermo chemical Conversion:

These processes under controlled temperature and oxygen conditions are used to convert the original biomass feedstock into more convenient forms of energy carriers, such as producer gas, oils or methanol. These carriers are either more energy dense and therefore reduce transport costs, or have more predictable and convenient combustion characteristics allowing them to be used in internal combustion engines and gas turbines.

- Pyrolysis—process of decomposing organic material with a limited supply of oxygen under high temperature and pressure.
- Carbonisation—Or Dry wood distillation is a pyrolyitc process optimized for the production of charcoal. Carbon accumulates due to a reduction in the levels of hydrogen and oxygen in the wood. The three basic types are 1) Internally heated, 2) Externally heated and 3) Hot circulating gas.
- Gasification—Under high temperatures and a controlled environment virtually all the raw material is being converted to gas. The biomass is partially combusted to form producer gas and charcoal. In the second stage, the CO_2 and H_2O produced in the first stage are chemically reduced by the charcoal, forming CO and H_2. These stages are spatially separated in the gasifier to trap the gases produced. Gasifier design will dependant on the feedstock characteristics. Gasification requires temperatures of about 800°C and is carried out in closed top or open top gasifiers. These gasifiers can be operated at atmospheric pressure or higher.
- Catalytic Liquefaction—Catalytic liquefaction is a low temperature, high pressure thermo chemical conversion process carried out in the liquid phase. It requires either a catalyst or a high hydrogen partial pressure. Progress is limited due to technical problems.

Biochemical conversion

A microbial electrolysis cell can be used to directly make hydrogen gas from plant matter As biomass is a natural material, many highly efficient biochemical processes have developed in nature to break down the molecules of which biomass is composed, and many of these biochemical conversion processes can be harnessed.

Biochemical conversion makes use of the enzymes of bacteria and other micro-organisms to break down biomass. In most cases micro-organisms are used to perform the conversion process: anaerobic digestion, fermentation and composting. Other chemical processes such as converting straight and waste vegetable oils into biodiesel is transesterification—a process during which an organic group in ester is replaced by an organic group of an alcohol.

Advantages

- **Carbon Neutral**—Biomass Energy results in no new net GHG emissions as it is part of the carbon cycle. Unlike coal and others forms of fossil fuel which have been buried millions of years ago and burning them adds to carbon in the atmosphere, responsible biomass energy generation results in no new carbon emissions or pollution
- **Use of Waste**—Biomass Energy is an Efficient Process which results in the use of mostly animal and crop waste which would be converted into carbon dioxide anyway. To use to as energy before the conversion is an important use of the millions of tons of waste that is generated by human activities
- **Continuous Source of Power**—Biomass Energy can be generated almost 24×7 unlike other forms of renewable energy like wind and solar that are intermittent in nature
- **Large variety of Feedstock**—Biomass Energy can use a large variety of feedstock such as wood pellet, rice husk, bagasse etc.
- **Low Capital Investment**—The capital investment required for biomass based power plants is relatively less at $1-2/watt of biomass capacity. Biomass Energy has similar capex characteristics as other forms of conventional energy like coal, gas, oil etc.
- **Distributed Nature**—Biomass plants can be built in remote areas and used as a distributed form of power generation
- **Can be built on small scale**—Biomass plants can be built in 1 Mw sizes unlike other types of power which require much larger scale. For example nuclear energy requires a typical plants size of at least 500 MW to make it economical
- **Reduces Methane which is a major GHG gas**—Decomposition of organic matter releases methane. Capturing this methane yields energy while protecting the atmosphere. The animal industry and landfills produce significant amounts of methane.
- **Easy Availability**—Biomass is readily available in almost all parts of the world except the desert areas where finding biomass is difficult
- **Low Cost Resource**—Biomass Energy can be produced cheaply if there is a good availability of feedstock

Disadvantages

- **Pollution in case of Poor Technology**—Biomass Energy can lead to air pollution in the form of char if the biomass is not completely combusted. This happens in the case of biomass energy being produced in rural areas through bad technology
- **Feedstock Problems**—One of the biggest drawbacks of biomass energy is the problem of feedstock. The plants are forced to run at lower utilization leading to higher costs if feedstock is not available due to some reason like a drought
- **Good Management Required**—The operations of a biomass plant requires very good management otherwise it may run into losses or even in some cases have to shut down. It requires a skill of high order to run the plant optimally and make use of alternative feedstock in case the regular one is not available
- **Limited Potential**—Biomass Energy has smaller potential than compared to other forms of energy like solar, hydro etc.
- **Controversial Plants**—Large Biomass Plants like the one in Scotland have run into massive protests as people think it might lead to air pollution and health hazards if constructed near their homes.

7) Geothermal Energy

Geothermal power is cost effective, reliable, sustainable, and environmentally friendly, but has historically been limited to areas near tectonic plate boundaries. Recent technological advances have dramatically expanded the range and size of viable resources, especially for applications such as home heating, opening a potential for widespread usage.

How it works

Geothermal energy is thermal energy stored in the Earth. Earth's geothermal energy originates from the original formation of the planet, from radioactive decay of minerals, from volcanic activity, and from solar energy absorbed at the surface. The geothermal gradient, which is the difference in temperature between the core of the planet and its surface, drives a continuous conduction of thermal energy in the form of heat from the core to the surface.

From hot springs, geothermal energy has been used for bathing since Paleolithic times and for space heating since ancient Roman times. Currently electricity is generated from geothermal energy worldwide and fed to the grid. The heat potential is used for district heating, space heating, spas, industrial processes, desalination and agricultural applications.

The Earth's geothermal resources are theoretically more than adequate to supply humanity's energy needs, but only a very small fraction may be profitably exploited.

Types / Classification

There are three types of geothermal steam plants depending on the way the energy is generated:

1. **Dry Steam Power Plant** produce energy directly from the steam generated underground. In this case we do not need additional boilers and boiler fuels because the steam (and no water) directly fill up the wells, passing through a rock catcher and directly operate the turbines. The using of such type is not popular because the natural dry steam hydrothermal reservoirs are very rare.

2. **Flash Steam Power Plant**—it is used when there is a liquid hydrothermal resource with high temperature (over 175°C). The operating principle is that when the hot water is released it is collected in a flash tank where the liquid is flashed to steam. The latter is separated from the liquid and it is used to run the turbines. The waste water is reinjected into the hydrothermal reservoir.

3. **Binary Steam Power Plant**—it is employed when the hydrothermal resource is with lower temperature (35°C). The hot water is passed to a heat exchanger where it is compound with secondary liquid with lower boiling point (hydrocarbon like isobutane or izopentane). This mixture vapor and its steam run the turbine. The waste mixture is recycled trough the heat exchanger. The geothermal fluid is condensed and it is returned to the hydrothermal resource. Since the most resources are with lower temperature the binary steam power plants are more common.

Current scenario

The International Geothermal Association (IGA) has reported that 10,715 megawatts (MW) of geothermal power in 24 countries is online, which is expected to generate 67,246 GWh of electricity in 2010. This represents a 20% increase in online capacity since 2005. IGA projects growth to 18,500 MW by 2015, due to the projects presently under consideration, often in areas previously assumed to have little exploitable resource.

Because geothermal power does not rely on variable sources of energy, unlike, for example, wind or solar, its capacity factor can be quite large—up to 96% has been demonstrated. The global average was 73% in 2005.

Installed Geothermal capacity—Top 10

Rank	Country	Installed capacity (GW)
1	USA	3
2	Philippines	1.9
3	Indonesia	1.2
4	Mexico	0.95
5	Italy	0.85
6	New Zealand	0.63
7	Iceland	0.58
8	Japan	0.54
9	El Salvador	0.2
10	Kenya	0.17

Advantages

- **Significant Cost Saving** : Geothermal energy generally involves low running costs since it saves 80% costs over fossil fuels and no fuel is used to generate the power.
- **Reduce Reliance on Fossil Fuels**: Dependence on fossil fuels decreases with the increase in the use of geothermal energy. With the sky-rocketing prices of oil, many countries are pushing companies to adopt these clean sources of energy.
- **Environmental Benefits**: Being the renewable source of energy, geothermal energy has helped in reducing global warming and

pollution. Moreover, Geothermal systems does not create any pollution as it releases some gases from deep within the earth which are not very harmful to the environment.
- **Direct Use:** Since ancient times, people having been using this source of energy for taking bath, heating homes, preparing food and today this is also used for direct heating of homes and offices.
- **Job Creation and Economic Benefits:** Geothermal energy on the other hand has created many jobs for the local people.

Disadvantages

- **Not Widespread Source of Energy:** Since, this type of energy is not widely used therefore the unavailability of equipment, staff infrastructure, training pose hindrance to the installation of geothermal plants across the globe.
- **High Installation Costs :** To get geothermal energy, requires installation of power plants, to get steam from deep within the earth and this require huge one time investment and require to hire a certified installer and skilled staff needs to be recruited and relocated to plant location
- **Can Run Out Of Steam :** Geothermal sites can run out of steam over a period of time due to drop in temperature or if too much water is injected to cool the rocks
- **Suited To Particular Region:** It is only suitable for regions where temperatures below the earth are quite low and can produce steam over a long period of time.
- **May Release Harmful Gases:** Geothermal sites may contain some poisonous gases and they can escape deep within the earth through the holes drilled by the constructors.

8) Tidal Energy

Tidal power, also called tidal energy, is a form of hydropower that converts the energy of tides into useful forms of power—mainly electricity.

Although not yet widely used, tidal power has potential for future electricity generation. Tides are more predictable than wind energy and solar power. Among sources of renewable energy, tidal power has traditionally suffered from relatively high cost and limited availability of sites with sufficiently high tidal ranges or flow velocities thus constricting its total availability. However, many recent technological developments have indicated that the economic and environmental costs may be brought down to competitive levels.

How it works

Tidal power is the only technology that draws on energy inherent in the orbital characteristics of the Earth-Moon system, and to a lesser extent in the Earth-Sun system. A tidal generator converts the energy of tidal flows into electricity. Because the Earth's tides are ultimately due to gravitational interaction with the Moon and Sun and the Earth's rotation, tidal power is practically inexhaustible and classified as a renewable energy resource. The daily rise and fall in the level of ocean water relative to the coastline is referred to as tide. Tides originate from the motions of the earth, moon and sun. The gravitational pull of the Moon and Sun along with the revolution of the Earth result in tides. (The magnitude of the gravitational attraction of an object is dependant upon the mass of an object and its distance.) The moon exerts a larger gravitational force on the earth, though it is much smaller in mass, because it is a lot closer than the sun. This force of attraction causes the oceans, which make up 71 percent of the earth's surface, to bulge along an axis pointing towards the moon. Tides are produced by the rotation of the earth beneath this bulge in its watery coating, resulting in the rhythmic rise and fall of coastal ocean levels. Coastal areas experience two high and two low tides over a period of 24 hours and slightly above.

Generating tidal energy

The technology required to convert tidal energy into electricity is comparable to technology used in traditional hydroelectric power plants. The first requirement is a dam across a tidal bay or estuary. However building a dam is expensive and the best sites are those where a bay has a narrow opening, thus reducing the length of dam required. Gates and turbines are installed. When there is adequate difference in the levels of the water on the different sides of the dam, the gates are opened. This causes water to flow through the turbines, turning the generator to produce electricity.

Electricity is generated by water flowing both inwards and outwards of a bay. There are periods of maximum generation every twelve hours, with no electricity generation at the six-hour mark in between. The turbines may also be used as pumps to pump extra water into the basin behind the dam at times when the demand on electricity is low. This water can later be released when the demand on the system is very high, thus allowing the tidal plant to function like a "pumped storage" hydroelectric facility.

Types / Classification

Tidal power can be classified into three generating methods:

Tidal stream generator

Tidal stream generators (or TSGs) make use of the kinetic energy of moving water to power turbines, in a similar way to wind turbines that use wind to power turbines.

Tidal barrage

Tidal barrages make use of the potential energy in the difference in height (or *head*) between high and low tides. Barrages are essentially dams across the full width of a tidal estuary. Tidal barrage work like a hydro-electric scheme, But that the dam is much bigger. A huge dam or a barrage is built across a river estuary. When the tide goes in and out, the water flows through tunnels in the dam.

As the tide comes in, the dam allows the sea water to pass through into a holding basin. As soon as the tide is about to go down, the dam is closed. The water held back in this way will be used to feed the turbine at low tide. The ebb and flow of the tides can be used to turn a turbine.

Dynamic tidal power

Dynamic tidal power (or DTP) is a theoretical generation technology that would exploit an interaction between potential and kinetic energies in tidal flows. It proposes that very long dams (for example: 30-50 km length) be built from coasts straight out into the sea or ocean, without enclosing an area. Tidal phase differences are introduced across the dam leading to a significant water-level differential in shallow coastal seas—featuring strong coast-parallel oscillating tidal currents such as found in the UK, China and Korea.

Advantages

- Apart from the initial installation cost, tidal power is free.
- Tidal energy produces no greenhouse gases or any other kind of pollution.

- It requires no fuel.
- Electricity is produced reliably.
- Not expensive to maintain.
- Tides are totally predictable, enabling us to calculate when we can generate more, and at times when the generation is low, shift the load to some other source of electricity generation.
- Offshore turbines and vertical-axis turbines are not extremely expensive to build and do not have a large environmental impact.

Disadvantages

- A barrage across an estuary is very expensive to build, and affects a very wide area.
- The environment is changed for many miles upstream and downstream.
- Many birds rely on the tide uncovering the mud flats so that they can feed. Fish can't migrate, unless "fish ladders" are installed.
- Tides provide power for only 10 hours each day, when the tide is actually moving in or out.
- There are only a few suitable sites for tidal barrages.

) Biofuel Energy

Biofuel is a type of fuel whose energy is derived from biological carbon fixation. Biofuels include fuels derived from biomass conversion, as well as solid biomass, liquid fuels and various bio gases. Although fossil fuels have their origin in ancient carbon fixation, they are not considered biofuels by the generally accepted definition because they contain carbon that has been "out" of the carbon cycle for a very long time. Biofuels are gaining increased public and scientific attention, driven by factors such as oil price spikes, the need for increased energy security, concern over greenhouse gas emissions from fossil fuels, and government subsidies.

How it works

Biofuels work on the principle that plants capture and store the sun's energy through the process of photosynthesis. Through a process of biocatalysis, the energy stored in the plant can be converted into any number of biofuel products—principally ethanol, biomethanol, biodiesel, biogas and biobutanol. The process of making each type of biofuel is different.

Types / Classification

First generation biofuels

'First-generation' or conventional biofuels are biofuels made from sugar, starch, and vegetable oil.

Bioalcohols

Biologically produced alcohols, most commonly ethanol are produced by the action of microorganisms and enzymes through the fermentation of sugars or starches (easiest), or cellulose (which is more difficult). Biobutanol (also called biogasoline) is often claimed to provide a direct replacement for gasoline, because it can be used directly in a gasoline engine (in a similar way to biodiesel in diesel engines).

Ethanol fuel is the most common biofuel worldwide, particularly in Brazil. Alcohol fuels are produced by fermentation of sugars derived from wheat, corn, sugar beets, sugar cane, molasses and any sugar or starch that alcoholic beverages can be made from (like potato and fruit waste, etc.). The ethanol production methods used are enzyme digestion (to release sugars from stored starches), fermentation of the sugars, distillation and drying. Ethanol can be used in petrol engines as a replacement for gasoline; it can be mixed with gasoline to any percentage. Most existing car petrol engines can run on blends of up to 15% bioethanol with petroleum/gasoline.

Methanol is currently produced from natural gas, a non-renewable fossil fuel. It can also be produced from biomass as bio methanol. The methanol economy is an interesting alternative to get to the hydrogen economy, compared to today's hydrogen production from natural gas. But this process is not state-of-the-art as the clean solar thermal energy process where hydrogen production is directly produced from water.

Butanol (C_4H_9OH) is formed by ABE fermentation (acetone, butanol, ethanol) and experimental modifications of the process show potentially high net energy gains with butanol as the only liquid product. Butanol will produce more energy and mostly can be burned "straight" in existing gasoline engines (without modification to the engine or car), and is less corrosive and less water soluble than ethanol, and could be distributed via existing infrastructures.

Biodiesel

Biodiesel is produced from oils or fats using transesterification and is a liquid similar in composition to fossil/mineral diesel. Transesterification is the process of exchanging the organic group R″ of an ester with the organic group R′ of an alcohol. Chemically, it consists mostly of Fatty Acid Methyl (or ethyl) Esters (FAMEs).

Feedstocks for biodiesel include animal fats, vegetable oils, rapeseed, jatropha, mustard, flax, sunflower, palm oil,etc. Pure biodiesel (B100) is the lowest emission diesel fuel. Biodiesel can be used in any diesel engine when mixed with mineral diesel. In some countries manufacturers cover their diesel engines under warranty for B100 use. B100 may become more viscous at lower temperatures, depending on the feedstock used. In most cases, biodiesel is compatible with diesel engines from 1994 onwards, which use 'Viton' (by DuPont) synthetic rubber in their mechanical fuel injection systems.

Green diesel

Green diesel, also known as renewable diesel, is a form of diesel fuel which is derived from renewable feedstock rather than the fossil feedstock used in most diesel fuels. Green diesel feedstock can be sourced from a variety of oils including canola, algae, jatropha and salicornia in addition to tallow. Green diesel uses traditional fractional distillation to process the oils, not to be confused with biodiesel which is chemically quite different and processed using transesterification.

Vegetable oil

Straight unmodified edible vegetable oil is generally not used as fuel, but lower quality oil can and has been used for this purpose. Used vegetable oil is increasingly being processed into biodiesel, or (more rarely) cleaned of water and particulates and used as a fuel.

To ensure that the fuel injectors atomize the vegetable oil in the correct pattern for efficient combustion, vegetable oil fuel must be heated to reduce its viscosity to that of diesel, either by electric coils or heat exchangers. This is easier in warm or temperate climates. Big corporations like MAN B&W Diesel, Wärtsilä, and Deutz AG as well as a number of smaller companies such as Elsbett offer engines that are compatible with straight vegetable oil, without the need for after-market modifications.

Vegetable oil can also be used in many older diesel engines that do not use common rail or unit injection electronic diesel injection systems. Due to the design of the combustion chambers in indirect injection engines, these are the best engines for use with vegetable oil. This system allows the relatively larger oil molecules more time to burn. Several companies like Elsbett or Wolf have developed professional conversion kits and successfully installed hundreds of them over the last decades.

Bioethers

Bio ethers (also referred to as fuel ethers or oxygenated fuels) are cost-effective compounds that act as octane rating enhancers. They also enhance engine performance, whilst significantly reducing engine wear and toxic exhaust emissions. Greatly reducing the amount of ground-level ozone, they contribute to the quality of the air we breathe.

Second generation biofuels (advanced biofuels)

Second generation biofules are biofuels produced from sustainable feedstock. Sustainability of a feedstock is defined among other by availability of the feedstock, impact on GHG emissions and impact on biodiversity and land use. Many second generation biofuels are under development such as Cellulosic ethanol, Alga fuel, biohydrogen, biomethanol, DMF, BioDME, Fischer-Tropsch diesel biohydrogen diesel, mixed alcohols and wood diesel.

Cellulosic ethanol production uses non-food crops or inedible waste products and does not divert food away from the animal or human food chain. Lignocellulose is the "woody" structural material of plants. This feedstock is abundant and diverse, and in some cases (like citrus peels or sawdust) it is in itself a significant disposal problem.

Current scenario

In 2010 worldwide biofuel production reached 105 billion liters (28 billion gallons US), up 17% from 2009, and biofuels provided 2.7% of the world's fuels for road transport, a contribution largely made up of ethanol and biodiesel. Global ethanol fuel production reached 86 billion liters (23 billion gallons US) in 2010, with the United States and Brazil as the world's top producers, accounting together for 90% of global production. The world's largest biodiesel producer is the European Union, accounting for 53% of all biodiesel production in 2010. As of 2011, mandates for blending biofuels exist in 31 countries at the national level and in 29 states/provinces. According to the International Energy Agency, biofuels have the potential to meet more than a quarter of world demand for transportation fuels by 2050.

Advantages of Biofuel

- The biofuel is the green friendly fuel which can help you to reduce the rising levels of green house caused by the oil.
- The biofuels can be made from different sources like plants, algae and fungi. All these sources are available in abundance therefore biofuel can be produced on massive scale.
- The biofuel will help us to reduce the dependence on OPEC and other foreign countries for energy needs
- The biofuels can help to reduce the global warming and it can also contribute in the global economy. The biofuels are much safer to handle as compared to gasoline. The health effects are also much lesser as compared to fossil fuels.

Disadvantages of Biofuel

- There might be some problem of carbon emission from machines which are used to cultivate and produce biofuels.
- Lots of cost required to produce biofuels on technological processes.
- There might be bad smell which is the outcome of the biofuel production cycle.
- There might be need to change the infrastructure of automobile to take the advantage of this natural fuel which requires lots of money.
- Biofuels production can at times require large amount of energy

0) Wave Energy:

Waves are a renewable source of energy that doesn't cause pollution. The energy from waves alone could supply the world's total electricity needs. The total power of waves breaking on the world's coastlines is estimated at 2 to 3 million megawatts. In some locations, the wave energy density can average 65 megawatts per mile of coastline. The problem is how to harness wave energy efficiently and with minimal environmental, social, and economic impacts.

How it works

Ocean waves are caused by the wind as it blows across the open expanse of water, the gravitational pull from the sun and moon, and changes in atmospheric pressure, earthquakes etc. Waves created by the wind are the most common waves and the waves relevant for most wave energy technology. Wave energy conversion takes advantage of the ocean waves caused primarily by the interaction of winds with the ocean surface. Wave energy is an irregular oscillating low-frequency energy source. They are a powerful source of energy, but are difficult to harness and convert into electricity in large quantities. The energy needs to be converted to a 60 Hertz frequency before it can be added to the electric utility grid.

Wave energy generation is a developing technology. Although many wave energy devices have been invented only a small number have been tested and evaluated and very few of these have been tested in ocean waves—testing is usually undertaken in a wave tank.

Types / Classification

There are three approaches to capturing wave energy:

- Floats or pitching devices
- Oscillating water columns
- Wave surge or focusing devices

Energy collection devices can be placed either on the shoreline near the shoreline, or offshore. Shoreline devices have the advantage of relatively easier maintenance and installation and do not require deep water moorings and long underwater electrical cables. The wave energy is less on the shoreline but this can be partly compensated by the concentration of wave energy that occurs naturally at some locations by refraction and/or diffraction.

Nearshore devices are situated in 10-25 metres of water near the shore. The most common device for this situation is the oscillating water column.

Offshore devices are situated in deep water, with typical depths of more than 40 metres. The incidence of wave power at deep ocean sites is three to eight times the wave power at adjacent coastal sites. A range of devices are being trialed for offshore use.

The wave energy converting device placed on the sea bed may be completely submerged, it may extend above the sea surface, or it may be a converter system placed on an offshore platform.

Current scenario

Despite inventors actively making systems to capture power from the waves, for the last two centuries, there is still not a wide application of wave energy devices as power generators. The availability of devices to fit different applications is not the problem—the technology is definitely there. The reality is that the only long term problem is making the technology work at a cost of power which a consumer is willing to pay. The system will work itself out. The price of fossil fuel generation will become more and more expensive and wave generated power will fall in price.

One of the biggest difficulties is in introducing a new, fledgling technology into a commercial market dominated by subsidised low cost fossil fuel and nuclear generation.

Advantages:

- the energy is free—no fuel is needed and no waste is produced
- not expensive to operate and maintain
- can produce a significant amount of energy.

Disadvantages:

- depends on the waves—variable energy supply
- needs a suitable site, where waves are consistently strong
- can disturb or disrupt marine life—including changes in the distribution and types of marine life near the shore

We have seen ten forms of renewable energy available for harnessing. All of them are in one way or other connected to Sun and its energy. **These wonderful gifts of nature to mankind provide us with many options of energy to accept and embrace for a sustainable living.**

CHAPTER 15

GREEN BUILDINGS

> *"In the end, our society will be defined not only by what we create, but by what we refuse to destroy."*
> —*John C. Sawhill*

> *"A green building is one which uses less water, optimises energy efficiency, conserves natural resources, generates less waste and provides healthier spaces for occupants, as compared to a conventional building."*
> Source: IGBC

What is a green building rating system?

A green building rating system or certification tool serves to provide a standard against which buildings with different levels of environmental design and efficiency can be compared. The primary objective of these mechanisms has been to stimulate market demand for buildings with improved environmental performance. An underlying premise is that if the market is provided with improved information and mechanisms, discerning clients can and will provide leadership in environmental responsibility and others will follow suit to remain competitive. Rating systems and labeling programs are considered one of the most potent and effective means to both improve the performance of buildings and transform market expectations and demand.

Reference: WGBC Publications

To explain green building concept let us look further.

Green building (also known as **green construction** or **sustainable building**) refers to a structure and using process that is environmentally responsible and resource-efficient throughout a building's life-cycle: from sitting to design, construction, operation, maintenance, renovation, and demolition. This practice expands and complements the classical building design concerns of economy, utility, durability, and comfort.

Although new technologies are constantly being developed to complement current practices in creating greener structures, the common objective is that green buildings are designed to reduce the overall impact of the built environment on human health and the natural environment by:

- Efficiently using energy, water, and other resources
- Protecting occupant health and improving employee productivity
- Reducing waste, pollution and environmental degradation

A similar concept is natural building, which is usually on a smaller scale and tends to focus on the use of natural materials that are available locally.

Green building practices aim to reduce the environmental impact of new buildings. The concept of sustainable development can be traced to the energy (especially fossil oil) crisis and the environment pollution concern in the 1970s. The green building movement in the U.S. originated from the need and desire for more energy efficient and environmentally friendly construction practices.

Environmental, economic, and social benefits are strong motives behind building green. However, modern sustainability initiatives call for an integrated and synergistic design to both new construction and in the retrofitting of an existing structure. Also known as sustainable design, this approach integrates the building life-cycle with each green practice employed with a design-purpose to create a synergy amongst the practices used.

Green building brings together a vast array of practices and techniques to reduce and ultimately eliminate the impacts of new buildings on the environment and human health. It often emphasizes taking advantage of renewable resources, e.g., using sunlight through passive solar, active solar, and photovoltaic techniques and using plants and trees through green roofs, rain gardens, and for reduction of rainwater run-off. Many other techniques, such as using packed gravel or permeable concrete instead of conventional concrete or asphalt to enhance replenishment of ground water, are used as well.

While the practices, or technologies, employed in green building are constantly evolving and may differ from region to region, there are fundamental principles that persist from which the method is derived:

- Sitting and Structure Design Efficiency
- Energy Efficiency
- Water Efficiency
- Materials Efficiency
- Indoor Environmental Quality Enhancement
- Operations and Maintenance Optimization
- Waste and Toxics Reduction

The essence of green building is an optimization of one or more of these principles. With the proper synergistic design, individual green building technologies may work together to produce a greater cumulative effect.

Sitting and structure design efficiency

The foundation of any construction project is rooted in the concept and design stages. The concept stage, in fact, is one of the major steps in a project life cycle, as it has the largest impact on cost and performance. In designing environmentally optimal buildings, the objective is to minimize the total environmental impact associated with all life-cycle stages of the building project. However, building as a process is not as streamlined as an industrial process, and varies from one building to the other, never repeating itself identically. In addition, buildings are much more complex products, composed of a multitude of materials and components each constituting various design variables to be decided at the design stage. A variation of every design variable may affect the environment during all the building's relevant life-cycle stages.

Energy efficiency

Green buildings often include measures to reduce energy use. To increase the efficiency of the building envelope, (the barrier between conditioned and unconditioned space), they may use high-efficiency windows and insulation in walls, ceilings, and floors.

Another strategy, passive solar building design, is often implemented in low-energy homes. Designers orient windows and walls and place awnings, porches, and trees to shade windows and roofs during the summer while maximizing solar gain in the winter. In addition, effective window placement (daylighting) can provide more natural light and lessen the need for electric lighting during the day. Solar water heating further reduces energy loads.

Onsite generation of renewable energy through solar power, wind power, hydro power, or biomass can significantly reduce the environmental impact of the building.

Water efficiency

Reducing water consumption and protecting water quality are key objectives in sustainable building. One critical issue of water consumption is that in many areas, the demands on the supplying aquifer exceed its ability to replenish itself. To the maximum extent feasible, facilities should increase their dependence on water that is collected, used, purified, and reused on-site. The protection and conservation of water throughout the life of a building may be accomplished by designing for dual plumbing that recycles water in toilet flushing. Waste-water may be minimized by utilizing water conserving fixtures such as ultra-low flush toilets and low-flow shower heads. Bidets help eliminate the use of toilet paper, reducing sewer traffic and increasing possibilities of re-using water on-site. Point of use water treatment and heating improves both water quality and energy efficiency while reducing the amount of water in circulation. The use of non-sewage and grey water for on-site use such as site-irrigation will minimize demands on the local aquifer.

Materials efficiency

Building materials typically considered to be 'green' include rapidly renewable plant materials like bamboo (because bamboo grows quickly) and straw, lumber from forests certified to be sustainably managed, ecology blocks, dimension stone, recycled stone, recycled metal, and other products that are non-toxic, reusable, renewable, and/or recyclable (e.g. Trass, Linoleum, sheep wool, panels made from paper flakes, compressed earth block, adobe, baked earth, rammed earth, clay, vermiculite, flax linen, sisal, seagrass, cork, expanded clay grains, coconut, wood fibre plates, calcium sand stone, concrete (high and ultra high performance, roman self-healing concrete), etc.) The EPA

(Environmental Protection Agency) also suggests using recycled industrial goods, such as coal combustion products, foundry sand, and demolition debris in construction projects. Building materials should be extracted and manufactured locally to the building site to minimize the energy embedded in their transportation. Where possible, building elements should be manufactured off-site and delivered to site, to maximise benefits of off-site manufacture including minimising waste, maximising recycling (because manufacture is in one location), high quality elements, better OHS management, less noise and dust.

Indoor environmental quality enhancement

The Indoor Environmental Quality (IEQ) category in LEED standards, one of the five environmental categories, was created to provide comfort, well-being, and productivity of occupants. The LEED IEQ category addresses design and construction guidelines especially indoor air quality (IAQ), thermal quality, and lighting quality.

Indoor Air Quality seeks to reduce volatile organic compounds, or VOC's, and other air impurities such as microbial contaminants. Buildings rely on a properly designed HVAC system to provide adequate ventilation and air filtration as well as isolate operations (kitchens, dry cleaners, etc.) from other occupancies. During the design and construction process choosing construction materials and interior finish products with zero or low emissions will improve IAQ. Many building materials and cleaning/maintenance products emit toxic gases, such as VOC's and formaldehyde. These gases can have a detrimental impact on occupants' health and productivity as well. Avoiding these products will increase a building's IEQ.

Personal temperature and airflow control over the HVAC system coupled with a properly designed building envelope will also aid in increasing a building's thermal quality. Creating a high performance luminous environment through the careful integration of natural and artificial light sources will improve on the lighting quality of a structure.

Operations and maintenance optimization

No matter how sustainable a building may have been in its design and construction, it can only remain so if it is operated responsibly and maintained properly. Ensuring operations and maintenance(O&M) personnel are part of the project's planning and development process will help retain the green criteria designed at the onset of the project. Every spect of green building is integrated into the O&M phase of a building's life. The addition of new green technologies also falls on the O&M staff. Although the goal of waste reduction may be applied during the design, construction and demolition phases of a building's life-cycle, it is in the O&M phase that green practices such as recycling and air quality enhancement take place.

Waste reduction

Green architecture also seeks to reduce waste of energy, water and materials used during construction. During the construction phase, one goal should be to reduce the amount of material going to landfills. Well-designed buildings also help reduce the amount of waste generated by the occupants as well, by providing on-site solutions such as compost bins to reduce matter going to landfills.

To reduce the impact on wells or water treatment plants, several options exist. "Greywater", wastewater from sources such as dishwashing or washing machines, can be used for subsurface irrigation, or if treated, for non-potable purposes, e.g., to flush toilets and wash cars. Rainwater collectors are used for similar purposes.

Centralized wastewater treatment systems can be costly and use a lot of energy. An alternative to this process is converting waste and wastewater into fertilizer, which avoids these costs and shows other benefits. By collecting human waste at the source and running it to a semi-centralized biogas plant with other biological waste, liquid fertilizer can be produced. This concept was demonstrated by a settlement in Lubeck Germany in the late 1990s. Practices like these provide soil with organic nutrients and create carbon sinks that remove carbon dioxide from the atmosphere, offsetting greenhouse gas emission. Producing artificial fertilizer is also more costly in energy than this process.

World Green buildings council:

The World GBC is a union of national Green Building Councils from around the world, making it the largest international organization influencing the green building marketplace.

Mission

To be the global voice for Green Building Councils and to facilitate the global transformation of the building industry towards sustainability. WGBC foster and support new and emerging Green Building Councils by providing them with the tools and strategies to establish strong organizations and leadership positions in their markets.

WGBC work closely with councils to advance their common interest by promoting local green building actions as solutions to address global issues such as climate change.

By driving collaboration between international bodies and increasing the profile of the green building market, WGBC work to ensure that green buildings are a part of any comprehensive strategy to deliver carbon emission reductions.

Ongoing projects to further the green building agenda include:

— World Green Building Day,
— Common Carbon Metrics project, and
— Collaborating with international bodies such as UNEP SBCI Sustainable Buildings Alliance (SBA) and the International Union of Architects (IUA).

WGBC showcase global best practice in green building through events and publications, such as the annual WorldGBC Congress, World Green Building Day and the WorldGBC Smart Market Report.

Since 1998, national Council representatives have met to review global activities and offer support for each other's efforts. This led to the founding meeting of the WorldGBC in November of 1999 in California, USA with 8 countries in attendance:

1. Australia
2. Canada
3. Japan
4. Spain
5. Russia
6. UnitedArabEmirates
7. UnitedKingdom
8. United States

Formal incorporation of the World Green Building Council (World) followed in 2002—its primary role being to formalize international communications, help industry leaders access emerging markets, and provide an international voice for green building initiatives.

Reference: World Green Building Council website

Indian Green Building Council (IGBC)

IGBC which is part of CII—Sohrabji Godrej Green Business Centre is actively involved in promoting the Green Building concept in India. The council is represented by all stakeholders of construction industry comprising of Corporate, Government & Nodal Agencies, Architects, Product manufacturers, Institutions, etc. The council operates on a consensus based approach and member-driven. The vision of the council is to usher green building revolution and India to become one of the world leaders in green buildings by 2015.

IGBC Services:

IGBC is facilitating the green building movement through the following services:

- Certification of Green Buildings in India
- IGBC Accredited Professional examination
- Green Building workshops & training programs
- Green Building missions
- Green Building Congress—India's flagship event on green buildings

Reference : Indian Green Building Council website

LEED® India

The Leadership in Energy and Environmental Design (LEED-INDIA) Green Building Rating System is a nationally and internationally accepted benchmark for the design, construction and operation of high performance green buildings.

LEED-INDIA provides building owners, architects, consultants, developers, facility managers and project managers the tools they need to design, construct and operate green buildings.

LEED-INDIA promotes a whole-building approach to sustainability by recognizing performance in the following five key areas:

- Sustainable site development
- Water savings
- Energy efficiency
- Materials selection and
- Indoor environmental quality

LEED-INDIA rating system provides a roadmap for measuring and documenting success for every building type and phase of a building lifecycle.

Specific LEED-INDIA programs include:

- LEED® India for New Construction (LEED® India NC)
- LEED® India for Core and Shell (LEED® India CS)

Refrence :Indian Green Building Council publications

IGBC Green Homes Rating System

Indian Green Building Council (IGBC) Green Homes is the first rating programme developed in India, exclusively for the residential sector. It is based on accepted energy and environmental principles and strikes a balance between known established practices and emerging concepts. The system is designed to be comprehensive in scope, yet simple in operation.

Benefits of Green Homes

A Green Home can have tremendous benefits, both tangible and intangible. The immediate and most tangible benefit is in the reduction in water and operating energy costs right from day one, during the entire life cycle of the building.

Tangible benefits

- Energy savings: 20-30 %
- Water savings: 30-50%

Intangible benefits

- Enhanced air quality,
- Excellent day lighting,
- Health & wellbeing of the occupants,
- Conservation of scarce national resources
- Enhance marketability for the project.

Eligibility

IGBC Green Homes Rating System is a measurement system designed for rating new residential buildings which include construction categories such as

- Individual homes
- High rise residential apartments,
- Gated communities
- Row houses
- Existing residential buildings which retrofit and redesigned in accordance with the IGBC Green Homes criteria.

Reference :Indian Green Building Council publications

CHAPTER 16

ELECTRIC VEHICLES

> *"Modern man does not experience himself as part of nature but as an outside force destined to dominate and conquer it. He even talks of a battle with nature, forgetting that if he won the battle, he would find himself in the losing side"*
>
> *E.F.Schumacher*

An electrical vehicle is a vehicle (which transports people from one place to another) that runs on electricity. The key difference with the fossil fuel powered vehicles is that it can be run using the electricity generated by fossil fuel and more importantly these electric vehicles can be run using energy generated from renewable sources. These vehicles can be categorized as zero emission vehicles. They have significantly lower negative impact on the environment than conventional vehicles. A unique feature of Electric or hybrid vehicles is regenerative braking system where they recover energy normally lost at the time of braking, and store them on onboard batteries. This in turn contributes to their overall efficiency. A hybrid electrical vehicle is a vehicle which combines conventional internal combustion engine with an electric powertrain.

A few interesting milestones in EV development

Period	Developments & Milestones
1828	Ányos Jedlik, a Hungarian who invented an early type of electric motor, created a tiny model car powered by his new motor.
1834	Vermont blacksmith Thomas Davenport, the inventor of the first American DC electrical motor, installed his motor in a small model car, which he operated on a short circular electrified track
1837	First locomotive built by Scotsman & Robert Davison of Aberdeenin
1879	First electric train was presented by Werner von Siemens at Berlin
1881	The world's first electric tram line opened in Lichterfelde near Berlin, Germany
1890	First successful working Railway underground line between City and South London in the UK was opened using electric locomotives built by Mather and Platt.
1891	An electric car in the conventional sense was developed in 1890 or 1891, by William Morrison of Des Moines, Iowa; the vehicle was a six-passenger wagon capable of reaching a speed of 14 miles per hour (23 km/h)
1895	The first use of electrification on a mainline was on a four-mile stretch of the Baltimore Belt Line of the Baltimore and Ohio Railroad (B&O)
1897	Electric vehicles found their first commercial application as a fleet of electrical New York City taxis, built by the Electric Carriage and Wagon Company of Philadelphia, was established

1899	The breaking of the 100 km/h (62 mph) speed barrier, by Camille Jenatzy in his 'rocket-shaped' vehicle Jamais Contente, which reached a top speed of 105.88 km/h (65.79 mph). Ferdinand Porsche's design and construction of an all-wheel drive electric car, powered by a motor in each hub, which also set several records in the hands of its owner E.W. Hart
1902	Italian railways were the first in the world to introduce electric traction for the entire length of a mainline the 106 km Valtellina line was opened
1917	The first gasoline-electric hybrid car was released by the Woods Motor Vehicle Company of Chicago. The hybrid was a commercial failure, proving to be too slow for its price, and too difficult to service.
1940's	Fuel rationing in United States caused Earle Williams to convert a motorcycle to electric power
1950's 1967	Henney Coachworks and the National Union Electric Company, makers of Exide batteries, formed a joint venture to produce a new electric car, the Henney Kilowatt. The car was produced in 36-volt and 72-volt configurations; the 72-volt models had a top speed approaching 96 km/h (60 mph) and could travel for nearly an hour on a single charge First Fuel Cell powered electric motorcycle created by Karl Kordesch at Union Carbide debuts
1971	An electric car received the unique distinction of becoming the first manned vehicle to be driven on the Moon; that car was the Lunar rover, which was first deployed during the Apollo 15 mission
1973	Mike Corbin sets first electric motorcycle land speed record of 101 mph.
1980s	Development of very high-speed service brought a revival of electrification. The Japanese Shinkansen and the French TGV were the first systems for which devoted high-speed electric railway lines were built from scratch.

1994	The REVA Electric Car Company was established in Bangalore, India, as a joint venture between the Maini Group India and AEV of California
1996	Peugeot Scoot'Elec was the first mass production of an electric motorbike.
2001	After seven years of research and development, it launched the REVAi, known as the G-Wiz i in the United Kingdom, in 2001
2000	The development of lithium-ion batteries and powerful electric motors (originally for military applications) made mainstream electric motorcycles more feasible
2006	A standard production Siemens Electric locomotive of the Eurosprinter type ES64-U4 (ÖBB Class 1216) achieved a speed of 357 km/h, the record for a locomotive-hauled train, on the new line between Ingolstadt and Nuremberg.
2008	California company Tesla Motors, for the electric sports car market, released the Lotus Elise-based Tesla Roadster
2009	2009 Zero Motorcycles hosted "The 24 Hours of Electricross", an all electric dirtbike race.
2010	ElectroCat was the first electric vehicle to set a record time in the Pikes Peak International Hill Climb The Nissan LEAF introduced in Japan and the United States is the first all electric, zero emission five door family hatchback to be produced for the mass market from a major manufacturer General motors released its offering the Chevy Volt
2011	Aptera Motors plan to release their futuristic 2e. Ford motors have also declared its commitment to deliver cars in large numbers where the customers will plug in to charge rather than refuel with fossil fuel.

The most common electrical vehicle is the lift or escalator seen around us even though these are not terms as such.
The electrical vehicles include

- Electric bicycle
- Electric scooters
- Electric cars
- Electric bus
- Electric train

Electrical bicycle:
An electric bicycle, also known as an e-bike, is a bicycle with an electric motor used to power the vehicle. Electric bicycles use rechargeable batteries and can travel with a speed of 15 to 20 miles per hour (24 to 32 km/h). Depending on the laws of the country in which they are sold, in some markets they are rapidly replacing traditional bikes and motorcycles.

The electric bicycle uses an electric motor to power the bicycle. There are direct drives as well as geared drives in the market. The pedal is power assisted by a chain drive, belt drive, hub motor or friction drives. Electric bicycles use rechargeable batteries to receive and store energy. Typical types used are Lead Acid, NiCd, NiMh etc Controllers these days electronically provide necessary power on off control to the bicycle. An electric bicycle can under normal terrain cover 50 to 70 Kms for one full charge

Electrical Scooters :
Electric motorcycles and scooters are vehicles with two or three wheels that use electric motors to attain locomotion.

These run on batteries charged through electricity. The batteries can be charged while parking and use this energy to power the vehicle to run through a motor. Different types electric scooters include two wheel medium speed, two wheel high speed,

Credit :Daniel.Cardenas at en.wikipedia

hree wheeled low, medium & high speed, free standing etc. The batteries used in there vehicles include lead acid, Lithium Iron, Lithium Polymer tc. The electric scooters can travel upto 80 Km in one full charging and an attain a speed of 80 Km/hr

There over a dozen electric bike manufactures/ assemblers in India. These include Hero group, Eko vehicle, Atlas, Yo—Bikes, BSA motors etc

Electric Car:

Electric vehicles first came into existence in the mid-19th century, when electricity was among the preferred methods for motor vehicle propulsion, providing a level of comfort and ease of operation that could not be achieved by the gasoline cars of the time. During early 20th century, the scarcity of electricity required for charging around the world and entry of a number of conglomerates into the oil business made electric cars take a dormant position.

Reva now Mahindra e2o, the world's only car to be produced consistently from 2001-02 from its plant in Bangalore—India will run up to 80 Km for a full charge. It is one of the best options for car owners who want to commute 40 to 45 Km on a daily basis.

In 2003, the first mass-produced hybrid gasoline-electric car, the Toyota Prius, was introduced worldwide, in the same year

Goin Green in London launched the G-Wiz electric car (Reva from Maini), a quadricycle that became the world's best selling EV, and the first battery electric car produced by a major auto company, the Nissan Leaf debuted in December 2010. Other major auto companies have electric cars in development, and various nations around the world are building pilot networks of charging stations to recharge them.

Lead acid batteries were earlier the batteries commonly used due to its ruggedness and availability. Lithium Ion batteries are the design focus of many in electric cars. These batteries have good energy density and high power. These cars will need periodic top up of the charge. Normally this can be done over night at home. This amount of energy input is key deterrent in the usage of these vehicles. The battery charging will take time, may be up to 8 hrs in some cases/models for full charge whereas fossil fuelled cars can have a full tank of fuel filled in a minute or so. However a lot of researches are being done to improve the time for charging.

The electric cars are generally more expensive due to the high cost of batteries. Of course the running and maintenance cost are minimal.

Electric Bus:

An electric bus is a bus powered by electricity. These are of different types. A bus may take about 2 to 3 tons of load. To further carry batteries to drive them is a difficult scheme. Hence the electric buses at many times adopt continuous / intermittent external charging. There are also buses under development that can hold the necessary energy within and run on routes away from chargers.

- The trolleybus is a type of electric bus powered by two overhead electric wires, with electricity being drawn from one wire and returned via the other wire, using two roof-mounted trolley poles
- The gapbus is a bus without rails or surface power lines, and it can share the road lane with other vehicles as well. Power is supplied over a gap of 12 cm (4.7 in) from a power line embedded in the ground.
- Capabus, runs without continuous overhead lines by using power stored in large onboard electric double-layer capacitors which are quickly recharged whenever the vehicle stops at any bus stop (under so called electric umbrellas), and fully charged in the terminus. These type of bus is under development in China.
- Gyrobus is an electric bus that uses flywheel energy storage not overhead wires like a trolleybus. It carries a large flywheel that is spun at up to 3,000 RPM by a "squirrel cage" motor. Fully charged, a gyrobus could typically travel as far as 6km on a level route at speeds of up to 50 to 60 km

Being a key mass rapid transport system, a lot of companies from different countries are working on bringing out a good electric bus. Foton American Bus company, Sinautec buses, Zonda Group, Jingsau Alfa bus company, Optare Uk are some of the companies offering electric bus.

Electric Train:

An electric multiple unit or EMU is a multiple unit train consisting of self-propelled carriages, using electricity as the motive power. An EMU requires no separate locomotive, as electric traction motors are incorporated within one or a number of the carriages. Most EMUs are used for passenger trains, but some have been built or converted for specialised non-passenger roles, such as carrying mail or luggage, or in departmental use, for example as de-icing trains. An EMU is usually formed of two or more semi-permanently coupled carriages, but electrically-powered single-unit railcars are also generally classed as EMUs.

EMUs are popular on commuter and suburban rail networks around the world due to their fast acceleration and pollution-free operation. Being quieter than DMUs and locomotive-drawn trains, EMUs can operate later at night and more frequently without disturbing residents living near the railway lines. In addition, tunnel design for EMU trains is simpler as provisions do not need to be made for diesel exhaust fumes.

The cars that form a complete EMU set can usually be separated by function into four types: power car, motor car, driving car, and trailer car. Each car can have more than one function, such as a motor-driving car or power-driving car.

- A power car carries the necessary equipment to draw power from the electrified infrastructure, such as pickup shoes for third rail systems and pantographs for over head systems, and transformers.
- Motor cars carry the traction motors to move the train, and are often combined with the power car to avoid high-voltage inter-car connections.
- Driving cars are similar to a cab car, containing a driver's cab for controlling the train. An EMU will usually have two driving cars at its outer ends.
- Trailer cars are any cars that carry little or no traction or power related equipment, and are similar to passenger cars in a locomotive-hauled train.

An electric locomotive is a locomotive powered by electricity from overhead lines, a third rail or an on-board energy storage device (such as a chemical battery or fuel cell). Electrically propelled locomotives with on-board fuelled prime movers, such as diesel engines or gas turbines, are classed as diesel-electric or gas turbine electric locomotives because the electric generator/motor combination only serve as a power transmission system. Electricity is used to eliminate smoke and take advantage of the high efficiency of electric motors.

One advantage of electrification is the lack of pollution from the locomotives themselves. Electrification also results in higher performance, lower maintenance costs and lower energy costs for electric locomotives.

Electric locomotives benefit from the high efficiency of electric motors, often above 90%. Additional efficiency can be gained from regenerative braking, which allows kinetic energy to be recovered during braking to put some power back on the line. Newer electric locomotives use AC motor-inverter drive systems that provide for regenerative braking

The most fundamental difference lies in the choice of direct (DC) or alternating current (AC). The earliest systems used direct current as initially, alternating current was not

Credit : Peter Van den Bossche

well understood and insulation material for high voltage lines was not available. Direct current locomotives typically run at relatively low voltage (600 to 3,000 volts); the equipment is therefore relatively massive because the currents involved are large in order to transmit sufficient power. Power must be supplied at frequent intervals as the high currents result in large transmission system losses.

As alternating current motors were developed, they became the predominant type, particularly on longer routes. High voltages (tens of thousands of volts) are used because this allows the use of low currents; transmission losses are proportional to the square of the current (e.g. twice the current means four times the loss). Thus, high power can be conducted over long distances on lighter and cheaper wires. Transformers in the locomotives transform this power to a low voltage and high current for the motors. A similar high voltage, low current system could not be employed with direct current locomotives because there is no easy way to do the voltage/current transformation for DC so efficiently as achieved by AC transformers.

Rectifier locomotives, which used AC power transmission and DC motors, were common, though DC commutators had problems both in starting and at low velocities. Today's advanced electric locomotives use brushless three-phase AC induction motors. These polyphase machines are powered from GTO-, IGCT- or IGBT-based inverters. The cost of electronic devices in a modern locomotive can be up to 50% of the total cost of the vehicle.

Electric traction allows the use of regenerative braking, in which the motors are used as brakes and become generators that transform the motion of the train into electrical power that is then fed back into the lines. This system is particularly advantageous in mountainous operations, as descending locomotives can produce a large portion of the power required for ascending trains.

In India the first electric train ran between Bombay's Victoria Terminus and Kurla along the Harbour Line of CR, on February 3, 1925, a distance of 9.5 miles. Following this a spate of electrification happened in the Mumbai suburbs. Then there was a long gap, and the next electrification project started only 1953 or 1954, in the Calcutta area (Howrah-Burdwan via Bandel, Sheoraphulli-Tarakeshwar), using 3kV DC traction. The first actual train run using 25kV AC was on December 15, 1959, on the Kendposi-Rajkharswan section (SER). India took the plunge from DC to AC electric traction in the mid-1950s. Since

French developments led the field, the AC locomotives supplied at first (from SNCF) followed that country's practice, whether built in India or France. These were the eight-wheeled WAM-1 locomotives that are still in operation in some places. Now Indian Railways has most of the bus routes of its network electrified (although not all), and this has resulted in about 65% of the traffic being hauled by electric traction.

Advantages and Disadvantages of Electric Vehicles

Advantages:

- No harmful emission thereby no air pollution
- Low noise pollution
- Higher efficiency results in 1/3 contribution of CO_2 emission from the source compared to IC engines
- Very low running cost and maintenance cost
- Electrical vehicle now does not need multiple gears to match power curves
- Very low or no vibration inside the moving vehicle
- Does not consume energy while Idling
- Cost of recharging is very less compared to fossil fuel engine vehicles
- Electrical vehicles are easy to drive and maneuver
- Electrical vehicles generate much lower amount of heat than IC engine vehicles

Disadvantages:

- Limited range—electrical vehicles have very limited range of operation now and will required very frequent charging
- It consumes lot of electricity and if not powered by renewables it will increase the use of coal or fossil fuel to generate the electricity—not good for environment
- Currently more expensive than fossil fuel vehicles
- Acceleration / Pick Up comparatively lower than IC engine powered vehicles
- Electrical vehicles need larger braking distances

Transportation accounts for over 30% of green house gas emission. This has a very serious impact on climate change and the future of our planet. The electric vehicles with almost zero emission can largely negate this impact. As technology improves for reducing the charging time and with cost of batteries coming down, we can hope that **one day we all will prefer to own and commute in electric vehicles with electricity harnessed fully from renewable energy.**

A list of 100 Solar Products

	Solar PV		Solar Thermal
Sl No.	Product Description	Sl No.	Product Description
1	Solar calculators	1	Solar water heater—Evacuated tube collector
2	Solar Lantern	2	Solar water heater—Flat plate collector
3	Solar Garden light	3	Solar water heater—pressurised
4	Solar LED flashlight	4	Solar dryer
5	Solar traffic light	5	Solar air conditioning
6	Solar home system _ Dc	6	Solar boiler
7	Solar home system—AC	7	Solar cooker
8	Solar bags	8	Solar furnace
9	Solar mobile charger	9	Solar air heater
10	Solar watch / clock	10	Solar water purifier
11	Solar agricultural water pumping	11	Solar disalination
12	Solar refrigerator	12	Solar steam generator
13	Solar chemical	13	Passive solar building
14	High Altitude Airship (HAA)	14	Urban heat island
15	Solar PV power station	15	Solar Pond
16	Solar Car	16	Solar baloons
17	Solar Boat	17	Solar sails
18	Solar camping lantern	18	Solar thermal power station
19	Solar Table light	19	Trombe Wall
20	Solar Key Chains	20	Stirling engine
21	Solar Wind Hybrid	21	Solar Absorption chillers
22	Solar Cap	22	Solar Oven
23	Solar search light	23	Solar micro CSP
24	Solar torch	24	Solar water heater—split system

25	Solar miner's hat	25	Solar water heater with heat exchanger
26	Solar flashlight	26	Solar thermal collector manifold
27	Solar traffic barrier light	**Solar PV Continued**	
28	Solar fencing		
29	Solar road stud	53	Solar gate opener
30	Solar radio	54	Solar address light
31	Solar courtyard light	55	Solar Electric Airconditioner
32	Solar laptop chargers	56	Solar remote survelliance
33	Solar arrow board	57	Solar sea navigator
34	Solar advertising board	58	Solar Valve
35	Solar multifunction charger	59	Solar back pack
36	Solar traffic display	60	Motion activated solar light
37	Solar fan	61	Solar landscape lighting
38	Solar plug and play system	62	Solar auto vent
39	Solar auto battery charger	63	Solar Pool Light
40	Solar toys	64	Solar learning kit
41	Solar Domestic water pump	65	Solar dock light
42	Solar fountain	66	Solar scout kit
43	Solar tile	67	Solar stepping stone
44	Solar mosquito repeller	68	Solar Bluetooth sound system
45	Solar wallet charger	69	Solar drier
46	Solar freezer	70	Solar Keyboard
47	Solar bicycle	71	Solar Mobile phone
48	Solar forest lights	72	Solar thermometer
49	Solar attic fan	73	Solar thermoelectric coolers
50	Daylighting	74	Solar Guitar tuner
51	Solar educational kit		
52	Solar parking lights		

Solar power plants around the world:

Solar Photovoltaic:

Sl No	Solar Power Station	Capacity (Mw)	Country	Developer/ Owner Companies
1	Sarnia Photovoltaic Power Plant	97	Canada	Enbridge, First Solar
2	Montalto di Castro Photovoltaic Power Station	84.2	Italy	Sunray, Sunpower
3	Finsterwalde Solar Park	80.7	Germany	Q cells, LDK solar
4	Rovigo Photovoltaic Power Plant	70	Italy	SunEdison, First Reseve corporation
5	Olmedilla Photovoltaic Park	60	Spain	Nobesol
6	Strasskirchen Solar Park	54	Germany	Q cells
7	Lieberose Photovoltaic Park	53	Germany	Juwi Group
8	Puertollano Photovoltaic Park	50	Spain	Renovalia,
9	Moura photovoltaic power station	46	Portugal	ACCIONA Energy
10	Kothen Solar Park	45	Germany	
11	Waldpolenz Solar Park	40	Germany	Juwi, First Solar
12	Ralsko Solar Park Ra 1	38.3	Czech Rep	
13	Copper Mountain Solar Facility	38	USA	Sempra Generation
14	Reckahn Solar Park	36	Germany	

15	Vepřek Solar Park	35.1	Czech Rep	Decci
16	Planta Solar La Magascona & La Magasquila	34.5	Spain	
17	Arnedo Solar Plant	34	Spain	T Solar
18	Planta Solar Dulcinea	31.8	Spain	Kyocera, Suntech
19	Tutow Solar Park	31	Germany	
20	Merida/Don Alvaro Solar Park	30	Spain	

Solar Thermal:

Sl No	Solar Power station	Capacity (Mw)	Country	Developer/Owner
1	Solar Energy Generating Systems	354	USA	Next Era Energy resources
2	Solnova	150	Spain	Abengoa Solar
3	Andasol solar power station	100	Spain	ACS Group (Andasol 1 & 2) Solar Millennium MAN Ferrostaal AG Stadtwerke München RWE Innogy
4	Extresol Solar Power Station	100	Spain	
5	Martin Next Generation Solar Energy Center	75	USA	Florida Power & Light Company (FPL)
6	Nevada Solar One	64	USA	Acciona Solar Power
7	Ibersol Ciudad Real	50	Spain	

8	Alvarado I	50	Spain	Acciona Energy
9	La Florida	50	Spain	
10	Majadas de Tiétar	50	Spain	
11	La Dehesa	50	Spain	
12	Palma del Rio 2	50	Spain	
13	Manchasol-1	50	Spain	
14	PS20 solar power tower	20	Spain	Abengoa Solar
15	Yazd integrated solar combined cycle power station	17	Iran	Doosan MAPNA
16	PS10 solar power tower	11	Spain	Abengoa Solar
17	Kimberlina Solar Thermal Energy Plant	5	USA	AREVA Solar
18	Sierra SunTower	5	USA	eSolar
19	Archimede solar power plant	5	Italy	Enel
20	Liddell Power Station Solar Steam Generator	2	Australia	

A list of Solar companies in India

Sl No	Company	Head Office Location
1	Tata Power Solar Systems	Bangalore
2	Bharath Heavy Electricals Limited	Bangalore
3	Kotak Urja	Bangalore
4	Emvee Solar	Bangalore
5	Anu Solar	Bangalore
6	Sudarshan Saur Shakthi	Aurangabad
7	Nuetech Solar systems	Bangalore
8	Orb Energy	Bangalore
9	Selco Solar	Bangalore
10	Velnet Energy Systems	Bangalore
11	Jain Solar	Jalgaon
12	Krishi Solar	Bangalore
11	V guard Industries	Kochi
13	Racold Thermo Limited	Pune
14	Photon Energy Systems	Hyderabad
15	Supreme Solar	Bangalore
16	Kamal Solar	Bangalore
17	Wipro Energy	Bangalore
18	Hykon Solar Energy	Thrissur
19	Reliance Solar	Mumbai
20	Moser Baer	New Delhi
21	Titan Energy Systems	Hyderabad
22	Sun Technics Energy Systems	Bangalore
23	Flareum Technologies	Mumbai
24	Greentek India private Limited	Secendrabad
25	Shriram Green Tech	New Delhi
26	Green Field Material handling	Thane
27	Hiramrut Energies	Rajkot

28	Patel Engineering	Rajkot
29	Yogi Solar Industries	Gonda
30	Electrotherm Renewable	Ahamedabad
31	Sun Energy Systems	Ananad
32	Suntron Energy	Gurgaon
33	Pure Solar Pvt Ltd	Kasauli—HP
34	Rashmi Solar	Bangalore
35	SunZone Solar	Bangalore
36	KraftSolar	Kochi
37	Soltrap Systems	Pune
38	Solarium Power Systems	Ludhiana
39	Illussions4real	Jaipur
40	Solsen Solar equipments	Madurai
41	Maharishi Solar Technology	Noida
42	Solace	Kolkata
43	Ammini Solar	Trivandrum
44	Velnet energy systems	Bangalore
45	Taylor Made Solar solutions	Valsad
46	Sharada Inventions	Nasik
47	Thermax Limited	Pune
48	Vijaya Industries	Udupi
49	Radiant Energy Technologies	Hyderabad
50	Goodsun Industries	Coimbatore

This list may not contain all the retail solar companies operating in India. The order in which it is listed does not reflect the size of the company or its recent performance.

Any omission of names is not intentional.

A List of Banks financing for Solar

1. Syndicate Bank
2. Canara Bank
3. Vijaya Bank
4. Bank Of India
5. Union Bank
6. Bank of Maharashtra
7. Karnataka Vikas Grammen Bank
8. Pragathi Grammen Bank
9. Punjab National Bank
10. Indian Bank
11. Andra Bank
12. Andra Pradesh Grameen Vikas Bank
13. Jharkand Grammen Bank
14. UCO Bank
15. Allahabad bank
16. Uttar Pradesh Grameen bank
17. Bank Of baroda
18. State bank of Saurashtra
19. Aryavart Gramin Bank
20. IREDA
21. State Bank of Mysore
22. Sarva UP Grameen Bank
23. J & K Grameen bank
24. Allahabad UP grammen bank
25. Maharashtra grameen bank

Domestic appliances: Power

Sl No	Description	Power (W)
1	Lights _Incandecent	40 to 100
2	Tube Lights	50 to 80
3	Lights _ CFL	5 to 18
4	Lights _ LED	1 to 5
5	Ceiling Fan	50 to 80
6	Television	120 to 200
7	Radio	50
8	Refrigerator	200 to 400
9	Electric Iron	1200 to 2400
10	Mixer	800 to 1200
11	Air Conditioner	2000 to 4000
12	Washing Machine	1500
13	Micro wave Oven	2000
14	Electric Grill	1800 to 2500
15	Toaster	1500 to 2000
16	Computer	200 to 350
17	Vaccum Cleaner	150 to 300
18	Electric Geyzer	1200 to 3000
19	Electric Hotplate	2000
20	Twin Door refrigerator	500 to 800

Units and Conversion Factors:

Watt is the unit of power
1 Watt is the rate at which the work is done or Joules per second
Watt = Voltage x Current
W = V x I
Energy is measured as power generated/ consumed for a time period

Watt hour is a unit of energy
W-Hr = W x Hr
1 kilowatt-hour is the energy of one kilowatt power flowing for one hour
Is commonly used as 1 Unit of electricity

Higher units of energy:
1,000 Watt Hr = 1 KiloWatt Hr = 1 Unit
1,000 KiloWatt Hr = 1 MegaWatt Hr
1,000 Mega Watt Hr = 1 Giga Watt Hr
1,000 Giga Watt Hr = 1 Tera Watt Hr
1,000 Tera Watt Hr = 1 Penta Watt Hr

Energy Conversion Table

	Kilo Joules	Kilo Watt hr	Kilo Calories	BTU	Tonne of oil Eq (TOE)
Kilo Joules	1	0.28 e-3	0.239	0.948	2.39e-8
Kilo Watt hr	3600	1	860	3412	8.6e-5
Kilo Calories	4.186	1.16e-3	1	3.97	1e-7
BTU	1.055	2.93e-4	0.252	1	2.52e-8
Tonne of oil Eq (TOE)	4.16e7	11630	1e7	3.97e7	1

Calorific Value

1 calorie of heat is the amount needed to raise 1 gram of water through 1 degree Centigrade at standard atmospheric pressure
1 calorie (cal) = 4.2 J
1 KWh = 3600 KJ = 860Kcal
1Kcal = 1.163 Wh

The calorific value of a fuel is the qty of heat produced by its combustion under standard test conditions.

Calorific Value of Various Fuels

S. No.	Name of Fuel	Unit	Calorific Value (kilocalories)
1.	Biogas	m3	4713
2.	Kerosene	kg	10638
3.	Firewood	kg	4500
4.	Cowdung cakes	kg	2100
5.	Coal	kg	4000
6.	Lignite	kg	2865
7.	Charcoal	kg	6930
8.	Soft coke	kg	6292
9.	LPG	kg	11500
10.	Furnace oil	kg	9041
11.	Coal gas	m3	4004
12.	Natural gas	m3	9000
13.	Electricity	kWh	860
14.	Diesel	kg	10700
15.	Petrol	kg	11300

ABOUT THE AUTHOR

K.K.Yadhunath is a renewable energy enthusiast based at Bengaluru (erstwhile Bangalore), India. Brought up in Thalassery—Kerala, did his graduation at Palghat, now working from Bengaluru for the last several years. He lives with his wife and two children.

He holds a degree in mechanical engineering and diploma in management. Working in renewable energy industry since 2004, currently at Orb Energy, Bengaluru as its Vice President—Design and Production, heading its mechanical design and manufacturing team.

He has two decades of professional experience in various industries including Renewable energy, Consumer appliances, Electronics etc He has worked in several organizations in India and Abroad handling responsible positions.

An avid Cricketer with interests including other sports, Aquarium, Arts etc. Passionate about scientific developments that make positive social impact, "Move Off The grid" is his first attempt as a professional author. With hands on experience in the subject and a great amount of research work, this wonderful collection is carefully written to provide excellent information that will make a good reading.

Email : yadhupost@gmail.com
Website : www.splendidscience.net
Facebook : Move Off The Grid

Note: The readers of this book is hereby informed that this book is the sole creation of the author and reflects his opinion and understanding, and that neither Orb Energy nor its associates endorse this book and will not be in any way responsible for any act or omission that may arise due to the publication of the book.